REMENT

OF

POWER SPECTRA

THE MEASUREMENT OF POWER SPECTRA

From the Point of View of Communications Engineering

By

R. B. BLACKMAN
Member of the Technical Staff
BELL TELEPHONE LABORATORIES, INC.

and

J. W. TUKEY

Member of the Technical Staff
BELL TELEPHONE LABORATORIES, INC.

and Professor of Mathematics
PRINCETON UNIVERSITY

DOVER PUBLICATIONS, INC.
NEW YORK

This new Dover edition, first published in 1959,
is an unabridged and corrected republication of
Part I and Part II of *The Measurement of
Power Spectra from the Point of View of Communications Engineering*, which originally ap-
peared in the January 1958 and March 1958
issues of Volume XXXVII of the *Bell System
Technical Journal*. The Index and Preface
were prepared especially for this edition by
the authors.

The work is published under the auspices and
through special arrangement with the *AMER-
ICAN TELEPHONE AND TELEGRAPH
COMPANY*.

Library of Congress Catalog Card Number: 59-10185

Manufactured in the United States of America.

Dover Publications, Inc.
180 Varick Street
New York 14, N. Y.

Table of Contents

i

Preface

If this account appears to be intended principally for communications engineers it is only because an adequate understanding of how (power*) spectrum analysis works seems to demand some aspects of a communications engineering approach. (Even some of our colleagues, interested in digital computation, or in statistical techniques, rather than in communications engineering, have reluctantly come to agree with this statement.) This account is intended for all who know *what* they want to accomplish by spectral measurement and analysis (though perhaps nothing of *how* to accomplish it), and are concerned with how to do it, or with how to think about doing it, or with why it should be done in one way rather than another. Considerable mathematical detail is given, but as a guide and background to practice, rather than either for its own sake or for the sake of rigor.

This account was written as an extended journal article, and not as an introduction to the beauties of spectral analysis. Thus, there is no discussion of why one might want to estimate (power) spectra, and no catalog of the wondrous results thus obtained. We cannot refrain, however, from quoting one wondrous result (from a letter from Walter E. Munk to one of the authors): ". . . we were able to discover in the general wave record a very weak low-frequency peak which would surely have escaped our attention without spectral analysis. This peak, it

---

* The use of the adjective "power" comes from a habit of language of communications engineers which we need not discuss here.

turns out, is almost certainly due to a swell from the Indian Ocean, 10,000 miles distant. Physical dimensions are: 1 mm high, a kilometer long."

Except rather summarily, later in this preface, we do not treat and distinguish the many sorts of Fourier methods which are available. All in all, our account has quite limited aims.

GROUPS OF READERS

We can, we feel, recognize four large groups of readers, as far as attitudes and backgrounds are concerned. There will be a group of communications engineers, to whom the sort of Fourier analysis we use will be relatively familiar, but whose statistical background may be quite limited. There will be a group of digital computermen whose clients have, or threaten to have, data on which spectral calculations will be required. Their background in the communications engineering approach is quite likely to be limited, as will be, perhaps, their background in statistics. There will be a group of statisticians without special background in this sort of time series analysis, to whom the communications engineering approach is very strange, and who wonder why "cosines have anything to do with things that are not periodic." There will, we hope, be a fourth group of readers who have, or know how to get, data they wish to analyze, and who may well lack background in all three fields: communications engineering, computer techniques, and statistics.

We say to the members of each group: "Though it may seem to you that it is not so, we have tried to consider you in writing this account. Matters are not simple, but we could have easily made them more complex." If we had been willing to try to write this account for some one group, or, more particularly, for some one part of one group, we should perhaps have been able to simplify its organization. It will be important for each reader to understand the structure actually used in this account, and to recognize that material complementary to that in Section x is to be sought in Section B.x. That when Section y expresses mainly statistical aspects, the communications engineering aspects are likely to be found in Section B.y. And *vice versa*. That, if Section z expresses the more formal aspects of a topic, the more verbal aspects are likely to be found in Section B.z. And *vice versa*.

Questions of Orientation

There are three sorts of questions which are likely to bother many readers at the beginning — questions to which answers are needed to set these readers on the right road.

(1) What are the different kinds of Fourier analysis? How are they to be distinguished? (What kinds of data are appropriately treated by these various methods? How can I tell which to use?)

(2) Why do cosines appear in the analysis of aperiodic phenomena? Isn't this unnatural?

(3) What other sorts of spectral analysis are there? How are they related to those discussed here?

We shall try to answer these questions briefly in the remainder of this preface.

The Different Kinds of Fourier Analysis

There are four ways of distinguishing among Fourier techniques, corresponding to questions which can be answered in a "yes or no" fashion. The first two are:

(1) Is the phenomenon treated periodic? (Or, equivalently, are all of the frequencies involved in the phenomenon integral multiples of a single frequency?)

(2) If the phenomenon is aperiodic, are the frequencies involved discrete (distinct) or continuous?

When we recognize that measured values will reflect not only the phenomenon studied but also measurement errors, and, usually, other sources of fluctuation, we see that the data will always be aperiodic. Thus, to the combination of these two questions there are but three essentially distinct answers, which may be phrased as follows:

(a) Phenomenon aperiodic, frequencies continuous, data aperiodic,

(b) Phenomenon aperiodic (but "almost periodic" in the mathematical sense), frequencies discrete but not harmonically related, data aperiodic.

(c) Phenomenon periodic, frequencies discrete and harmonically related, data aperiodic.

A third way of distinguishing among Fourier techniques corresponds to the question

(3) Are the times involved discrete (and equi-spaced) or continuous? As explained in Sections 13 and 14 for one case, the practical distinction between discrete and continuous time is negligible (although there may be differences of mathematical theory). The answer to this question is usually not important.

A fourth way of distinguishing among Fourier techniques corresponds to the question

(4) Are the data analyzed thought of as unique, or as a statistical sample?

The answer to this question is exceedingly important, even though it might not seem that this should be the case.

Either of the answers to (4) can be combined with answer (a), or (b), or (c) to (1) and (2), making six possible such combinations. We have thus six pairs of kinds of Fourier analysis, the two kinds in a pair differing only in whether time is continuous or discrete.

In three of these pairs we wish to treat the data as a statistical sample. The text discusses in detail the pair corresponding to answer (a) to questions (1) and (2).

No especial development of theory or practice seems to have yet occurred for the other two statistical-sample pairs. Data which one might possibly suppose to fall under one of these pairs are usually analyzed by the methods described in the text, just as if they fell under the first pair. This practice makes more sense when we realize that no finite amount of data can determine whether a function is aperiodic, almost periodic, or periodic. A seemingly aperiodic function *might* start to repeat exactly with a period of, say, three times the length of the available record. A seemingly periodic function may suddenly cease repeating at a point just outside the available record. And the inevitable concealment of fine detail by measurement noise (by fluctuations and errors of measurement) blurs such distinctions even more completely.

There remain three pairs of kinds of Fourier analysis, in all of which we wish to treat the given data as coming from a unique function. Probably the simplest pair involves periodic phenomena, arising, for example, in connection with seasonal (or daily) fluctuations in climate, weather, or economic series, or in connection with physical phenomena related to an angular orientation (in a plane). In practice a very important distinction must be made between cases in which

(i) the periodic phenomenon dominates the data, irregularities (fluctuations or errors) of measurement or recording being negligible,

(ii) the periodic phenomenon must be sought among substantial irregularities, and

(iii) the periodic phenomenon is to be separated from both irregularities and slow trends.

In the uncomplicated case (i), almost any method of analysis will be satisfactory if its arithmetic is kept simple. Such methods are discussed, often under the heading of "empirical harmonic analysis", in books on numerical computation (especially in Theodore R. Running's *Empirical Formulas*).

Methods of analysis for case (ii) may have to be more subtle. The outstanding reference is Chapter XVI, "Periodicities and Harmonic Analysis in Geophysics" by Julius Bartels on pp. 545–605 (Volume 2) of *Geomagnetism* by Sidney Chapman and Julius Bartels, Oxford University Press (1940, second edition 1951). (Consideration should also be given to the techniques developed by the Labroustes; see references 41, 42, 43 in the text.)

In case (ii) situations, the use of methods appropriate to case (i) situations may be very dangerous.

For case (iii) we have a choice. Most case (ii) methods are applicable with but little change and provide relatively reliable answers. Rather specific methods are widely used in economics, where usually, and probably wisely, the periodic parts are not expressed in Fourier form. Such methods are discussed in many texts in economic statistics.

(The most complete summary of the classical Fourier series and periodogram approaches seems to be that of Karl Stumpf, *Grundlagen und Methoden der Periodenforschung*, Springer, Berlin, 1937; Edwards, Ann Arbor, 1945, who also provided convenient tables, *Tafeln und Aufgaben zur harmonischen Analyse und Periodogrammrechnung*, Springer, Berlin, 1939. Note that Stumpf's first graph, on page 45 of *Grundlagen und Methoden der Periodenforschung*, illustrates aliasing. A number of related topics are treated in Cornelius Lanczos, *Applied Analysis*, Prentice-Hall, Englewood Cliffs, 1956).

The next pair involves almost periodic phenomena, which are of frequent occurrence in geophysics (lunar and solar tides, etc.) and occurs, as an approximation, in such astronomical problems as variable-star light curves. For the *rare* instances in which the periodic terms are well separated in intensity *and* the irregular background is small, excellent use can be made of the treatment in Whittaker and Robinson's *Calculus of Observations*. The best reference, however, is the Bartels chapter already mentioned.

There remains one pair of kinds of Fourier analysis, in which the data is thought of as unique and aperiodic. The only natural excuse for bringing cosines into such situations is a desire to study the effect of one or more linear time-invariant transformations. For a theoretical problem, the theory and practice of Fourier integrals stand ready at hand. For a data analysis problem, very little help is available, since the Fourier integral transform always involves more parameters than we can hope to estimate from any finite amount of data. In practice, then, we must approximate this pair of kinds of analysis by some other pair. If we treat the data as a sample rather than unique, we come to the pair treated in the text. If we treat the phenomenon as periodic, or almost periodic, rather than aperiodic, we come to one of the pairs just discussed above. Some such shift is essential.

In practical data analysis, of course, we must always be prepared to approximate wisely. A pressure wave representing human speech, for example, is not precisely periodic, but many short stretches of it are so nearly periodic that a periodic analysis is often very fruitful.

Equally important with a willingness to approximate (far enough, but not too far) is a willingness to take a statistical view whenever the data represents at most "something like" the situation of eventual concern. (It is indeed very rare in practice that the data under analysis represent the identical situation about which conclusions are to be drawn.)

WHY COSINES FOR APERIODIC PHENOMENA?

It is surely natural, once a statistical view is taken, to study averages, first of linear expressions and then of quadratic expressions. It is also natural to begin by studying situations which are independent of clock-setting in the sense that whatever started to happen at one (clock) time could equally likely have started at any other (clock) time. (Such situations are termed stationary, in technical language.) The average values of all linear expressions can then be described by one number (because the average values of all individual values are all the same).

Average values of quadratic expressions are more complicated. To describe them we have to specify, not just a single number, but a single function. (If time is discrete, this function can often be specified by an infinite sequence of numbers.) There are many choices of such *bases*, each of which will serve to describe the average values of all quadratic expressions. All bases contain the same information, but some code it in more secret (hard to read) ways than others. The power spectrum itself, the autocorrelation function (or, better, the autocovariance function), and the average values of $[X(t + \tau) - X(t)]^2$ (as a function of τ), are only three among many choices. *Exact* and *complete* knowledge of the values of any choice, of course, determines the values of each of the others. However, approximate knowledge of the power spectrum does not determine the others very well, nor does approximate knowledge of either function of lag determine the power spectrum very well. Frequency and time correspond to very different forms of coding — small, obvious errors in terms of either one may correspond to a nearly nonsensical message in terms of the other.

It is natural for the statistician to say "So what, I wanted to use the autocovariance function anyway. All that you have told me merely encourages me to forget cosines and power spectra." To this view, there are two counter-arguments of importance, one general, the other statistical.

Whether it be called a stochastic process, a noise, or a signal, we are very likely to want to know what will happen when it is subjected to a time-invariant linear transformation; when, otherwise expressed, it is passed through a linear, time-invariant filter. If we are thinking of it statistically, we are almost certainly concerned with averages of simple expressions for the output. (The natural basis for the averages of linear expressions, the average value at any time, will, of course, be multiplied by the voltage transfer function of the filter, evaluated at zero frequency.) The output values, expressed in terms of any basis for the averages of quadratic expressions, will be linearly expressible in terms of the corresponding input values, and may be regarded as the result of a linear transformation on the values describing the input. However, some linear transformations are simpler than others. Surely the diagonal transformations are the simplest. The transformation expressing the effect of any filter at all, as seen in terms of the values appearing in the *power spectrum* description, is diagonal. The transformations corresponding to the various lag functions are almost never diagonal, no matter how simple the filter.

Thus, if we desire convenience in handling average values of quadratic (or general second-degree) expressions, (i) we will use some basis purely for compactness, (ii) among all bases we will select the power spectrum basis because of its much simpler transformation properties. Simplicity alone forces us to consider cosines in connection with aperiodic stochastic processes. This is the general reason.

The statistical reason lies in the joint behavior of sampling fluctuations. Suitably spaced estimates of (smoothed) power spectral density fluctuate in a rather thoroughly independent manner. On the other hand, estimates of functions of lag (be they autocorrelations, autocovariances, or average values of $[X(t + \tau) - X(t)]^2$) have fluctuations which are so far from independence as to frequently fool almost anyone who examines tables or graphs of their values. From this point of view, also, the power spectrum has important advantages. All in all, there is little hope of escaping cosines.

RELATED TYPES OF SPECTRAL ANALYSIS

Estimation of spectra of single functions or of single series (which is the only subject discussed in the text) is not the only form of spectral analysis. When two or more concurrent series are available (originating from related phenomena), cross-spectra can be calculated, two for each pair of series. (The theory has been discussed by Goodman, Ref. 1 in the text.) Such analyses provide opportunities for studying such problems as: the behavior of certain linear transformations (by studying cross-spectra between input and output); the two-dimensional structure of atmospheric turbulence (by studying cross-spectra between vertical and horizontal components of wind velocity); the direction of arrival of ocean waves (by studying cross-spectra of records at different locations); and the gust structure of the atmosphere well away from the ground (by using cross-spectra between airplane accelerations and control positions to determine what these accelerations would have been with the controls locked). Besides the rather technical account by Goodman, brief introductions are given by Press and Tukey (cited in the bibliography) and in a paper by Tukey "An Introduction to the Measurement of Noise Spectra" which appears in *Surveys in Probability and Statistics*, Almquist and Wiksell, Stockholm, 1958.

February 1958.
Bell Telephone Laboratories, Inc.,
    Murray Hill, New Jersey,
       and
Center for Advanced Study in the Behavioral Sciences, Inc.,
    Stanford, California.

THE MEASUREMENT
OF
POWER SPECTRA

PART I

General Discussion

I. INTRODUCTION

Communications systems and data-processing systems are generally required to handle a large variety of signals in the presence of noise. The design of these systems depends to a large extent upon the statistical properties of both the signals and the noise. In most cases, the noises may be represented, or approximated, as stationary Gaussian random processes with zero averages, so that all of their relevant statistical properties will be contained by the autocovariance function or the power spectrum. In many cases, the signals may also be represented, or approximated, as stationary Gaussian random processes with zero averages.

Noises, signals, or other ensembles of functions (given continuously or at intervals) which are approximately stationary but not Gaussian are often also usefully studied in terms of autocovariance functions or power spectra. Although the average and the spectrum are no longer the *only* relevant statistical properties, they are usually the most useful ones. Thus, we shall do well to keep as much of our treatment generally applicable as possible.

In almost every case, the autocovariance function or power spectrum of either the noise or the signal will be of interest and importance.

To determine the autocovariance function or power spectrum of an (approximately) stationary random process, we are often reduced to the necessity of measurement and computation. Exact determination would require a perfectly-measured, infinitely-long piece of a random function (or a collection of pieces of infinite total length), and would require infinitely detailed computations. Both of these requirements are, of course, impractical. Approximate determination, on the other hand, raises the questions of how much data of a given accuracy will be required, what computational approach should be used, and how much reliance may be placed upon the results. Practically useful answers to these questions may be found by combining results from transmission theory and the theory of statistical estimation. These answers prove to be relatively simple. The only major difficulty in their practical application is the extensiveness of the data required for highly precise estimates. This requirement is an inherent, irrevocable characteristic of such random processes.

In this account we shall treat only the measurement of spectra of individual noises or signals. The measurement and utilization of the cross-spectra of pairs of series is also important, but is beyond our present scope. Questions of distribution and anticipated variability of cross-spectral estimates, and of certain estimates derived from them, have recently been cleared up by Goodman.[1]

It is natural to feel that the measurement of power spectra is simple, and that no problems deserving extended discussion arise. After all, are there not commercial "wave analyzers" of many sorts; have not Fourier series served for many years to analyze the frequencies of many signals, (musical instruments, human voices, etc.)? Why should there be a serious problem?

There are two reasons why elementary methods fail us rather frequently. On the one hand, the signal may not be available in indefinitely long time stretches. Either the conditions of observation, experimental or otherwise, or the difficulties of careful recording may make it impractical to have so much data that we can afford to analyze carelessly. (The examples of Sections 26 to 28, involving spectra of radar tracking, noise in very short-lived devices, and irregularities in the earth's rotation, respectively, all illustrate this point). Even if observation and recording can be afforded, the cost of computation often forces careful analysis.

On the other hand, the *random* nature of much noise, and some signals, in which the relative amplitudes and phases of different frequencies are not stably related (in contrast to voices and musical notes), introduces much more difficulty with sampling fluctuations and provides much more significant appearing, thus much more misleading statistical artefacts than experience with simpler signals would lead investigators to expect. (In postwar oceanography, for example, high mechanical ingenuity was expended in the construction of simple and effective wave analysers to produce detailed spectra of ocean waves. The results were quite misleading, because the frequency resolution obtained was too high for the limited length of records used, and almost the entire appearance of the resulting spectra was an illusion due to the particular fluctuations of the particular record. The use of broader filters has since led to meaningful results which could be related to physically satisfying theories.) All too often, the practical study of spectra requires care.

Effective measurement of power spectra requires understanding of a number of considerations and action guided by all of them. Explaining each individual consideration is necessary, but it is equally necessary to explain how they fit together. The general structure of this description of spectral measurement is the following: an introduction to the concepts (Sections 1–3), brief accounts of individual considerations (Sections 4–19), accounts of how these considerations are assembled in analysis (Sections 20–21), and planning for measurement (Sections 22–28, which include discussion of examples), and Sections in Part II giving the details supporting the earlier sections.

We have attempted to provide, somewhere, most of the facts and attitudes that are needed in the practical analysis of (single) power spectra.

Readers interested in either completing their present knowledge or in gaining a brief overview of the subject may wish to proceed next to Sections 20ff, whence they can be referred to specific sections of interest. For some, reading of Sections 1–3 may be a helpful preliminary for Sections 20ff. For others, who want to build more solidly as they go, reading straight through, perhaps with considerable cross-reference to Part II, may be best.

A function of time $X(t)$ generated by a random (or stochastic) process is one of an ensemble of random functions which might be generated by the process. The value of the function at any particular point in time is thus a random variable with a probability distribution induced by the ensemble. Furthermore, the values of the function at any particular set of points, say $t = t_1, t_2, \cdots, t_n$, have an n-dimensional joint probability distribution also induced by the ensemble. Such probability distributions have an important bearing on the design of any communication system or data-processing system which must handle an output from such a random process, be this output "signal" or "noise".

We shall often, but not always, assume that the random process is *Gaussian*. This means that, for every n, $t_1, t_2, \cdots, t_n$, the joint probability distribution of

$$X(t_1), \; X(t_2), \; \cdots, \; X(t_n),$$

is an n-dimensional Gaussian or normal distribution. Each such distribution is completely determined by the ensemble *averages*

$$\bar{X}(t_i) = \text{ave} \, \{X(t_i)\},$$

and by the *covariances*

$$C_{ij} = \text{cov} \, \{X(t_i), X(t_j)\}$$
$$= \text{ave} \, \{[X(t_i) - \bar{X}(t_i)] \cdot [X(t_j) - \bar{X}(t_j)]\} \, .$$

As a matter of convenience in development we will assume that the averages $\bar{X}(t_i)$ are zero. The covariances then reduce to

$$C_{ij} = \text{ave} \, \{X(t_i) \cdot X(t_j)\}.$$

Throughout, we will assume that the random process is *stationary* (that is, temporally homogeneous) in the sense that it is unaffected by translations of the origin for time. The covariances C_{ij} now depend only on the time separation $t_i - t_j$ so that

$$C_{ij} = C(t_i - t_j).$$

Thus, the noise is completely specified by a single function of a single variable. In particular, $C(0)$ is the *variance* (for zero average, the average square) of $X(t)$.

If the process were stationary, with zero averages, but were not Gaussian, then knowledge of the covariance as a function of lag, although providing a very large amount of useful information, would not completely specify the process. The results of this paper fall into two categories: (i) those relating to average values of spectral estimates, and (ii) those relating to variability of spectral estimates. The average-value results apply generally under the assumptions of stationarity (and zero averages), and do not depend upon the Gaussian assumption. The variability results are exact under the Gaussian assumption, and are usually rather good approximations otherwise. Thus, our results have practical value for noises and signals which are not closely Gaussian.

Results about variability are naturally used: (i) for planning the approximate extent of measurement effort, (ii) for indicating the presence of changes, during a series of measurements, in the quantities estimated, and (iii) as a means of judging the precision of an over-all estimate. The results given here are mainly for the first planning use. The additional uncertainties in actual variability due to either non-normality of distribution, or to changing of conditions between runs, or to both, are often all too real, but are rarely large enough to affect planning seriously. The same is true of mild nonstationarity. The Gaussian, stationary results can also be applied to the second use, the detection of changes in the true spectrum, but considerable caution is in order. The precision of final over-all values is ordinarily far more wisely judged from the observed consistency of repeated measurements (as by analysis of variance of logarithms of various spectral density estimates at the same nominal frequency) than from any theoretical variability based on a Gaussian assumption.

Communications engineers are more accustomed to work with a single time function of infinite extent than with an ensemble of finite pieces (of such functions). It is perhaps fortunate, therefore, that averages across an ensemble are equivalent (ergodicity) to averages over time along a single function of infinite extent, whenever a process is stationary, Gaussian, has zero averages, and has a continuous power spectrum (no "lines"). (If the process were not stationary the single function approach could not be used in this way.)

Since we seek to make this account as intuitive as possible for communications engineers, we shall define transforms, and make many other computations in terms of averages along a single function (as limits of integrals over centered intervals). In dealing with more specifically statistical issues, however, we shall write "ave" for average value, "var" for variance and "cov" for covariance, and shall do nothing to hinder the interpretation of these operators as acting across the ensemble. (Those who wish can also think of them in single function terms.)

The covariance at lag τ, in single function terms, is given by

$$C(\tau) = \lim_{T \to \infty} \frac{1}{T} \int_{-T/2}^{T/2} X(t) \cdot X(t + \tau) \cdot dt.$$

In ensemble terms, we would write merely

$$C(\tau) = \text{ave} \{X(t) \cdot X(t + \tau)\}.$$

The function $C(\tau)$ is frequently called the autocorrelation function, although historical usage in both statistics and the theory of turbulence (Taylor[2]) shows that this name should be applied to the (normalized) ratio $C(\tau)/C(0)$. We shall call $C(\tau)$ the *autocovariance* function. This name is appropriate to our formal definition of $C(\tau)$ because we have assumed that the averages of our process are all zero. Whenever we give up the assumption of zero averages, as we must almost always do when dealing with actual data, we shall use ave $\{X(t) \cdot X(t + \tau)\}$ = average lagged product,

ave $\{[X(t) - \bar{X}] \cdot [X(t + \tau) - \bar{X}]\}$ = autocovariance function,

where $\bar{X}$ is the common value of ave $\{X(t)\}$ and ave $\{X(t + \tau)\}$, thus preserving accurate usage.

Because of the direct relationship of the joint probability distribution to the autocovariance function, much of the statistical attention given to Gaussian stationary time series (time-sampled random functions) has been expressed in terms of serial-correlation coefficients (corresponding to lag-sampled autocovariance functions).

A stationary Gaussian random process may be regarded (e.g. Rice[3]) as the result of passing *white* Gaussian noise through a fixed linear network with a prescribed transmission function. White Gaussian noise, in turn, may be regarded as the superposition of the outputs of a set of simple harmonic oscillators (continuously infinite in number) with

(a) a continuous distribution in frequency,*

(b) uniform amplitude over the significant frequency range of the transmission system, and

(c) independent and random phases.

This point of view is particularly suited to the techniques employed by communications engineers. It is therefore not surprising that communications engineers have dealt with stationary Gaussian random processes almost entirely in terms of power spectra.

Because the autocovariance function and the power spectrum are Fourier transforms of each other, it would at first appear to be purely a matter of convenience which one is used in any particular situation. Indeed, optimum filter characteristics for the protection of signal against noise in communications systems and in many types of computing devices have, on occasion, been determined by the use of the autocovariance function. In practice, however, the filter designer almost invariably turns to the power spectrum as the final criterion of adequate design and performance.

In practice also, where the autocovariance function or the power spectrum must be determined by measurement and computation, and then interpreted, the choice is now heavily weighted in favor of interpretation of the power spectrum. Although a great deal of theoretical work has been done on the probability distribution of the serial-correlation coefficients for Gaussian stationary time series of finite length, with a view to the estimation of the confidence which may be placed upon practical

---

* The term "frequency" is used throughout this paper in the communications engineer's sense, viz., cycles per second of a sinusoidal wave. (Exceptional uses in the statistician's sense are explicitly noted.)

results, the criteria which have been developed along this line are so complicated that it is extremely difficult to apply them in practice, where the joint distribution must be considered. On the other hand, the situation with respect to the power spectrum is now very satisfactory for practical purposes. This stems from results obtained by Tukey,[4] and in part independently by Bartlett,[5] about nine years ago, when studies were made of the effects of sampling, of finite length of series, and of choice of computational procedure on the behavior of the estimated power spectrum. Since that time, applications to such diverse fields as ocean waves (Marks and Pierson[6]), aerodynamics (Press and Houbolt[7]), meteorology (Panofsky[8]), and seismology (Wadsworth, Robinson, Bryan, and Hurley[9]), have shown the practical applicability of these results to a wide variety of physical time series.

Shortly after these studies first reached the stage of practical usefulness, the theoretical analysis was reformulated by Blackman, who expressed it from the point of view of transmission theory, for presentation to members of the technical staff of Bell Telephone Laboratories (Out-of-Hours Courses 1950–1951, Communications Development Training Program 1950–1952).

More recent contributions (1950–1957) to the theory of power spectrum estimation have been reviewed by Bartlett and Medhi,[10] by Bartlett,[11] and by Grenander and Rosenblatt.[12]

2. AUTOCOVARIANCE FUNCTIONS AND POWER SPECTRA

First, let us consider the ideal case. The autocovariance function which was defined in the preceding section by

$$C(\tau) = \lim_{T \to \infty} \frac{1}{T} \int_{-T/2}^{T/2} X(t) \cdot X(t + \tau) \cdot dt$$

may be reduced to the form

$$C(\tau) = \int_{-\infty}^{\infty} P(f) \cdot e^{i2\pi f\tau} \, df$$

where

$$P(f) = \lim_{T \to \infty} \frac{1}{T} \left| \int_{-T/2}^{T/2} X(t) \cdot e^{-i2\pi ft} \, dt \right|^2$$

(cp. Section B.2). The function of frequency $P(f)$ describes the *power spectrum* of the stationary random process considered. More precisely $P(f) \, df$ represents the contribution to the *variance* from frequencies between f and $(f + df)$. If we think of $X(t)$ as a voltage across (or current through) a pure resistance of one ohm, the long-time average power dissipated in the resistance will be strictly proportional to the variance of $X(t)$. This important special case is the excuse for the adjective "power". The pure statistician might prefer to refer to the covariance spectrum or to the second moment spectrum rather than to the power spectrum. For precision, we shall often refer to $P(f)$ as the *spectral density* or *power spectral density*. When no confusion is likely, we may call $P(f)$ merely the *power spectrum*.

The relation exhibiting the autocovariance function as the Fourier transform of the power spectrum may be inverted to express the power spectrum as the Fourier transform of the autocovariance function. Thus, we have

$$P(f) = \int_{-\infty}^{\infty} C(\tau) \cdot e^{-i2\pi f \tau} \, d\tau \,.$$

The autocovariance function $C(\tau)$ and the power spectrum $P(f)$ are, formally at least, even functions of their respective arguments. Hence, the relation between them may be expressed more simply as two-sided cosine transforms, viz.

$$C(\tau) = \int_{-\infty}^{\infty} P(f) \cdot \cos 2\pi f \tau \cdot df,$$

and

$$P(f) = \int_{-\infty}^{\infty} C(\tau) \cdot \cos 2\pi f \tau \cdot d\tau;$$

or perhaps even more simply, as one-sided cosine transforms, viz.

$$C(\tau) = 2 \int_{0}^{\infty} P(f) \cdot \cos 2\pi f \tau \cdot df$$

and

$$P(f) = 2 \int_{0}^{\infty} C(\tau) \cdot \cos 2\pi f \tau \cdot d\tau.$$

Results are usually more conveniently developed in terms of the two-sided forms than in terms of the one-sided forms. In Sections A.3 and B.4 for example, the use of the two-sided forms with exponential kernels will be found to simplify considerably the expression of the operation of convolution between functions of lag or of frequency. In Section B.6, the use of the two-sided forms with exponential kernels avoids some complicated manipulations of trigonometric identities in the early stages of the development.

It should be particularly noted that

$$\text{ave } \{X(t)\cdot X(t + \tau)\} = \int_0^\infty 2P(f)\cdot\cos 2\pi f\tau\cdot df$$

and that (setting $\tau = 0$)

$$\text{var } \{X(t)\} = \int_0^\infty 2P(f)\cdot df.$$

Thus, it is evident that our definition of the power spectrum differs from the usual one which associates the power spectrum only with positive frequencies. References to the power spectrum in practice are usually in terms of a density $2P(f)$ over $0 \leqq f < \infty$ only.

3. THE PRACTICAL SITUATION

In practice we can obtain only a limited number of pieces of $X(t)$ of finite length. Each piece may be regarded as a sample drawn from a population or universe of pieces of $X(t)$ of the same length. The reduction of the data will therefore yield no more than estimates of the auto-covariance function and of the power spectrum — estimates which are subject to sampling variations and to biases in the usual statistical sense. This situation is further complicated in those cases in which we can measure, or desire to use, only values of $X(t)$ at uniformly spaced values of t within each piece of $X(t)$; in other words, those cases in which we are dealing with classical time series (discrete time) rather than with time functions (continuous time).

The theoretical study of sampling variability and bias is much simpler in the case of the estimates of the power spectrum than in the case of the estimates of the autocovariance function (or of serial-correlation co-efficients). This reflects the fact that, as we consider longer and longer records, and two narrow frequency bands with an arbitrarily small but fixed separation, we may find estimates of the power in these frequency bands which both (i) become arbitrarily precise, and (ii) become arbitrarily nearly (statistically) independent. The existence of such estimates is another particular consequence of the Gaussian character, as expressible in terms of "random and independent phases", of the random process from which we have one or more samples.

Use of the power spectrum has an additional advantage over use of the autocovariance function. In almost all practical situations, the data analyzed does not represent the actual output of the random process. In such cases the data will have been modified, appreciably if not radically, by the transmission characteristics of the devices employed to secure the data. This modification of the data may in fact be intentional, as we shall see when we come to the discussion of "prewhitening" in Section 15. In any case, the estimates will have to be corrected for the effects of this modification of the data. For estimates of the power spectrum, the correction procedure is a simple division of a frequency function by another frequency function. For estimates of the autocovariance function, however, the correction procedure will require a Fourier transformation, division of the resulting frequency function by another frequency function, and an inverse Fourier transformation. This whole sequence of operations on the autocovariance function is the only practical procedure for the inversion of the convolution (see Section A.3) which is the effect to be corrected for. (Details are given at the end of Section B.3.)

As we shall see, the measurements and computational operations may involve the use of either analog or digital computation and handling of either continuous "signals" or discrete data. (Whatever be its relation to some communication or data-handling system, we shall call continuous-time signals or noise which we are analyzing "signals", while discrete-time signals or noise, or discrete-time samples thereof, will be called data.) In actual practice, and for well-defined reasons of instrumentation and computation engineering, only a few of the many possible combinations are used.

Spectrum analysis by analog computation is almost always applied to continuous "signals", and makes use of filtering rather than going through autocovariance or mean lagged products. Digital computation must be carried out on discrete data, perhaps time-sampled from a continuous "signal", and preferably uses an indirect route via mean lagged products rather than trying to isolate individual frequency bands directly. In either case, each data value must enter several computations, and it is rarely economic to carry these computations out directly in real time, especially since there will not usually be enough such analysis on a regular basis to saturate the working capacity of the analog or digital computer used. Consequently, recording, either of "signals" or of data or of both, is almost inevitable.

Thus, five stages will be important in nearly every case:
  (1) sensing (pick-up, conversion, etc.)
  (2) transmission (to recorder or, possibly, to computer)
  (3) recording (including play-back, and, perhaps, time-sampling)
  (4) computation (formulas, computing circuit performance, etc.)
  (5) interpretation.
In every one of these stages, quality of performance (noise level, distortion, etc.) will be of importance.

The present account concentrates on the computational and interpretational stages, but indicates, from time to time, those considerations in the other stages which are peculiar to power spectrum analysis.

We have been unable to find a wholly satisfactory arrangement for the material we wish to present. In order to facilitate a relatively easy once-over, these introductory sections now continue into a condensed account, from which proofs, some reasons, and many helpful remarks have been postponed to the Appendix and sections in Part II. Readers interested in a survey may find it adequate to read only the condensed account. Others may find it best to skim this condensed account first, to read Appendix A next, and then to study similarly numbered sections of Part II and the condensed account.

The continuous record of finite length will be treated first (Sections 4–11); the modifications required for the discrete equally spaced record are covered next (Sections 12–21), and the opening account concludes with a discussion of the planning and analysis of measurement programs (Sections 22–28).

Appendix A (Sections A.1 to A.6) treats fundamental Fourier techniques, and the transform-pairs most closely associated with diffraction, in both the continuous and equi-spaced cases.

Each section of Part II relates to the similarly numbered section of the main body, and contains details of derivations, further reasons, and additional helpful remarks.

An Index of Notations is given on pages 161-65, a Glossary of Terms on pages 167-78, and acknowledgements are made on page 178.

CONTINUOUS RECORDS OF FINITE LENGTH

4. FUNDAMENTALS

Given a continuous record of finite length, it is clear that we cannot estimate the autocovariance function $C(\tau)$ for arbitrarily long lags. Surely, no estimate can be made for lags longer than the record. Furthermore, as we will find in due course, it is usually not desirable to use lags longer than a moderate fraction (perhaps 5 or 10 per cent) of the length of the record. Thus, in place of

$$C(\tau) = \lim_{T \to \infty} \frac{1}{T} \int_{-T/2}^{T/2} X(t) \cdot X(t + \tau) \cdot dt$$

for all values of τ, we will have at our disposal

$$C_{00}(\tau) = \frac{1}{T_n - |\tau|} \int_{-(T_n - |\tau|)/2}^{(T_n - |\tau|)/2} X\left(t - \frac{\tau}{2}\right) \cdot X\left(t + \frac{\tau}{2}\right) \cdot dt = C_{00}(-\tau)$$

only for $|\tau| \leqq T_m < T_n$, where T_n is the length of the record, and T_m is the maximum lag which we desire to use. We will call $C_{00}(\tau)$ the *apparent autocovariance function*, since (on account of ergodicity) its average value is $C(\tau)$ for $|\tau| \leqq T_m$.

The class of estimates for the power spectrum with which we are chiefly concerned will be derived from a *modified apparent autocovariance function* by Fourier transformation. While the modified apparent autocovariance functions, which are obtained by multiplying the apparent autocovariance function by suitable even functions of τ, are often far from being respectable estimates of the true autocovariance function, their transforms are very respectable estimates of *smoothed* values of the true spectral density.

Let $D_i(\tau)$ be a prescribed even function of τ, subject to the restrictions $D_i(0) = 1$, and $D_i(\tau) = 0$ for $|\tau| > T_m$, (where $i = 0, 1, 2, 3, 4$, depending upon the shape of $D_i(\tau)$ for $|\tau| < T_m$), and let the corresponding modified apparent autocovariance function be defined by

$$C_i(\tau) = D_i(\tau) \cdot C_{00}(\tau).$$

We may regard $D_i(\tau)$ as a window of variable transmission which modifies the values of $C_{00}(\tau)$ differently for different lags. It is therefore natural to call $D_i(\tau)$ a *lag window*.

For any lag window which meets the conditions stated above, $C_i(\tau)$ is calculable from the data. Further, it is clear that $C_i(\tau) = 0$ for $|\tau| > T_m$ although $C_{00}(\tau)$ was not defined there. Because $C_i(\tau)$ is defined for all values of τ, it has a perfectly definite Fourier transform $P_i(f)$, which should satisfy the symbolic relation,

$$P_i(f) = Q_i(f) * P_{00}(f)$$

where $Q_i(f)$ is the Fourier transform of $D_i(\tau)$, the asterisk indicates convolution (see Section A.3 for discussion), and $P_{00}(f)$ is the Fourier transform of $C_{00}(\tau)$. However, $P_{00}(f)$ is not determinate because $C_{00}(\tau)$ is not specified for $|\tau| > T_m$ (and its definition cannot be directly extended beyond $|\tau| = T_n$). Nevertheless, since

$$\text{ave} \{C_i(\tau)\} = D_i(\tau) \cdot C(\tau)$$

where $C(\tau)$ is the true autocovariance function, it follows that

$$\text{ave} \{P_i(f)\} = Q_i(f) * P(f)$$

where $P(f)$ is the true power spectrum, that is, the Fourier transform of $C(\tau)$. The average may be thought of as either across the ensemble, or along time. (The latter type of averaging would correspond to replacing

$X(t)$ by $X(t - \lambda)$, thus changing the stretch of $X(t)$ which is observed, and then averaging over λ.) The corresponding explicit relation, viz.

$$\text{ave } \{P_i(f_1)\} = \int_{-\infty}^{\infty} Q_i(f_1 - f) \cdot P(f) \cdot df$$

exhibits the average value of $P_i(f_1)$ as a smoothing (average-over-frequency) of the true power spectrum density $P(f)$ over frequencies "near" f_1 with weights proportional to $Q_i(f_1 - f)$. In a manner of speaking $P_i(f_1)$ is the collected impression of the true power spectrum $P(f)$ obtained through a window of variable transmission $Q_i(f_1 - f)$. It is therefore natural to call $Q_i(f)$ the *spectral window* corresponding to the *lag window* $D_i(\tau)$.

The form just given for ave $\{P_i(f_1)\}$ is natural for our two-sided definition of power spectra, but, in order to view the result from the standpoint of transmission theory for real-valued signals, it is convenient to express the result in a form appropriate to a one-sided definition of power spectra. Taking advantage of the fact that $Q_i(f)$ and $P(f)$ are even functions, we may write

$$\text{ave } \{2P_i(f_1)\} = \int_0^{\infty} H_i(f; f_1) \cdot 2P(f) \cdot df$$

where

$$H_i(f; f_1) = Q_i(f + f_1) + Q_i(f - f_1)$$

and where we recall that $2P(f) \, df$ is the amount of power between f and $(f + df)$ in the one-sided true power spectrum. Similarly, $2P_i(f) \, df$ is the amount of power between f and $(f + df)$ in the one-sided estimated power spectrum. The function $H_i(f; f_1)$ has one of the necessary properties of a physically realizable power transfer function inasmuch as it is an even function of f as well as of f_1. In general, however, it does not have the property of being non-negative at all frequencies f. Nevertheless, it is a convenient function to use in the analysis of the variability of the estimated power spectrum. It will be convenient to regard the average value of the smoothed power density estimate ave $\{2P_i(f_1)\}$ as the result of passing the true power spectrum, through a "network" with *power transfer function* $H_i(f; f_1)$.

We see that our procedures will lead us to estimates whose average values are a smoothing (average-over-frequency) of the true power spectral density $P(f)$ over frequencies "near" f_1, and not to estimates of $P(f_1)$ itself. The problem of choosing the shape of the lag window $D_i(\tau)$ so that its Fourier transform $Q_i(f)$ will be concentrated near $f = 0$ is almost identical to the problem of choosing an intensity distribution along an antenna so that most of the radiation from the antenna will fall in a narrow beam. From this analogy we will use such terms as *main lobe* and *side lobes* for the principal maximum and subsidiary extrema of $Q_i(f)$. (Indeed, any attempt to confine the power transfer function to too narrow a frequency band — too narrow in comparison with the reciprocal of the longest lag used — would be analogous to an attempt to construct a practical hyperdirective antenna.)

It is not surprising that we are led to estimate a smoothed power spectrum. With only a finite length of $X(t)$ available, we should not expect to be able to identify frequencies exactly, and are, indeed, unable to do so. (The presence of neighboring frequencies with random phases will have effects similar to those of noise in preventing such identification.)

5. TWO PARTICULAR WINDOW PAIRS

In order to specify a particular family of estimates within the class of estimates defined in the preceding section, we have only to specify $D_i(\tau)$ or $Q_i(f)$. We would like to concentrate the main lobe of $Q_i(f)$ near $f = 0$, keeping the side lobes as low as feasible. In order to concentrate the main lobe we have to make $D_i(\tau)$ flat and rather blocky. In order to reduce the side lobes, however, we have to make $D_i(\tau)$ smooth and gently changing. Since $D_i(\tau)$ must vanish for $|\tau| > T_m$ we must compromise. So far, cut-and-try inquiry has been more powerful in finding good compromises than has any particular theory.

A simple and convenient compromise is represented by the lag window (whose use is called "hanning")

$$D_2(\tau) = \frac{1}{2}\left(1 + \cos\frac{\pi\tau}{T_m}\right) \quad \text{for} \quad |\tau| < T_m$$

$$= 0 \quad \text{for} \quad |\tau| > T_m.$$

(Window pairs 0 and 1 are discussed in Section B.5.) An alternative compromise is represented by the lag window (whose use is called "hamming")

$$D_3(\tau) = 0.54 + 0.46\cos\frac{\pi\tau}{T_m} \quad \text{for} \quad |\tau| < T_m$$

$$= 0 \quad \text{for} \quad |\tau| > T_m.$$

These lag windows and the corresponding spectral windows are illustrated in Fig. 1. Notice that the main lobes are four times as wide as the

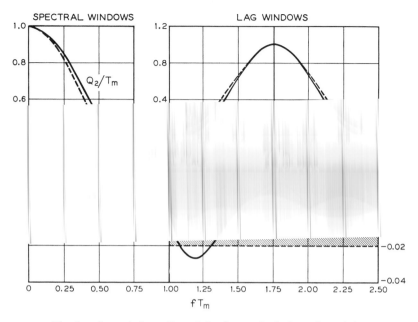

Fig. 1 — Lag windows D_2 and D_3. Spectral windows Q_2 and Q_3.

side lobes (excepting the split side lobes nearest the main lobes), and that the (normal) side lobe width is $1/(2T_m)$.

The general nature of the spectral windows in these two pairs is the same: a main lobe, side lobes at most 1 per cent or 2 per cent of the height of the main lobe. There are differences, which are sometimes relevant, but these may not be obvious. The two most important of these differences are:

(a) The highest side lobe for the "hamming" (spectral) window is about $\frac{1}{3}$ the height of the highest side lobe for the "hanning" window,

(b) The heights of the side lobes for the "hanning" window fall off more rapidly than do those for the "hamming" window.

One difference favors one pair, and one the other.

These and several other window pairs are discussed in Section B.5.

6. COVARIABILITY OF ESTIMATES — BASIC RESULT

It is shown in Section B.6 that, strictly only under Gaussian circumstances, the covariance of any two power density estimates of the sort we have been considering is given to a good degree of approximation by

$$\text{cov}\ \{2P_i(f_1),\ 2P_j(f_2)\} \approx \int_0^\infty H_i(f;f_1)\cdot H_j(f;f_2)\cdot 2\Gamma(f)\cdot df$$

where the *power-variance spectrum* $\Gamma(f)$ depends only on the true power spectrum $P(f)$ and the *effective record length* T_n, as described below. Thus, we may regard the covariance of the two power density estimates as the result of passing the power variance spectrum $\Gamma(f)$ through two networks in tandem, one with power transfer function $H_i(f;f_1)$, the other with power transfer function $H_j(f;f_2)$. In other words, we may regard the covariance (of the estimates of the power spectrum) as the power remaining from the *power-variance spectrum* $\Gamma(f)$ after passing through the *two* windows $H_i(f;f_1)$ and $H_j(f;f_2)$ associated with the estimates themselves. *If the windows do not overlap, the estimates do not covary* (at least not in terms of second moments).

In particular, of course,

$$\text{var}\ \{2P_i(f_1)\} = \text{cov}\ \{2P_i(f_1),\ 2P_i(f_1)\}$$

$$\approx \int_0^\infty H_i(f;f_1)^2\cdot 2\Gamma(f)\cdot df$$

to which we can give a similar interpretation.

These results would become exact if we were to replace $C_{00}(\tau)$ by

$$\tilde{C}_{00}(\tau) = \frac{1}{T_n - T_m}\int_{-(T_n-T_m)/2}^{(T_n-T_m)/2} X\left(t - \frac{\tau}{2}\right)\cdot X\left(t + \frac{\tau}{2}\right)\cdot dt,$$

where $|\ \tau\ | \leqq T_m < T_n$. In $C_{00}(\tau)$ we averaged $X(t - (\tau/2))\cdot X(t + (\tau/2))$ over an interval of t of length $T_n - |\ \tau\ |$, varying with τ. In $\tilde{C}_{00}(\tau)$ we would be averaging $X(t - (\tau/2))\cdot X(t + (\tau/2))$ over an interval of t of length $T_n - T_m$ independent of τ. We could actually do this because $|\ t \pm (\tau/2)\ | \leqq T_n/2$ for $|\ \tau\ | \leqq T_m$. However, for values of $|\ \tau\ |$ less than T_m, $\tilde{C}_{00}(\tau)$ would not make use of some values of $X(t - (\tau/2))\cdot X(t + (\tau/2))$ which are used in $C_{00}(\tau)$. Thus, $\tilde{C}_{00}(\tau)$ would be wasteful. It seems best, therefore, to use $C_{00}(\tau)$ for computation, but to approximate its variability by the variability corresponding to a $C'_{00}(\tau)$ which could not be calculated from the actual values. This "approximate" hypothetical $C'_{00}(\tau)$ involves a fixed range of integration T'_n part way between $T_n - T_m$ and T_n. The situation is illustrated in Fig. 2, where the ranges of integration are shown for the actually "feasible" $\tilde{C}_{00}(\tau)$, for the $C_{00}(\tau)$ which "wastes not", and for the $C'_{00}(\tau)$ which we use to "approximate" $C_{00}(\tau)$. The shaded areas delineate the products which are actually available.

The best choice of an intermediate value depends somewhat upon the $D_i(\tau)$ and $D_j(\tau)$ involved, and is discussed in Section B.6. In practically useful cases we may take

$$T'_n = T_n - \tfrac{1}{3}T_m .$$

The power-variance spectrum is given approximately and closely by

...g... ...ich piece they came from, then we may use this formula for $\Gamma(f)$ with

$$T'_n = T_n - \frac{p}{3} T_m$$

as a satisfactory approximation for the *effective total length*.

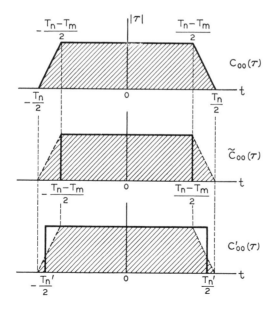

Fig. 2 — Range of integration over t as a function of τ in $\tilde{C}_{00}(\tau)$, $C_{00}(\tau)$, and $C'_{00}(\tau)$.

7. COVARIABILITY OF ESTIMATES — APPROXIMATE FORMS

In assessing the covariability of estimates of the smoothed power spectrum, the relative magnitudes of three distances along the frequency axis are important:

(a) the distance $1/T'_n$, the reciprocal of the effective length of record,

(b) the least distance over which $P(f)$ changes by an important amount for f near f_1, and

(c) the least distance over which $H_i(f; f_1)$ changes by an important amount for f near f_1 (this is of the order of $1/T_m$ and is usually much larger than $1/T'_n$).

If $P(f)$ changes slowly enough to make (b) larger than (a), we may use the approximation

$$\Gamma(f) \approx \frac{2}{T'_n} [P(f)]^2$$

whence, approximately,

$$\text{cov } \{P_i(f_1), P_j(f_2)\} \approx \frac{1}{T'_n} \int_0^\infty P_{i1}(f) \cdot P_{j2}(f) \cdot df$$

where

$$P_{i1}(f) = H_i(f; f_1) P(f)$$
$$P_{j2}(f) = H_j(f; f_2) P(f).$$

In the same terms we have

$$\text{ave } \{P_i(f_1)\} = \int_0^\infty P_{i1}(f) \cdot df$$

and

$$\text{ave } \{P_j(f_2)\} = \int_0^\infty P_{j2}(f) \cdot df.$$

The relation of covariances to averages thus established may be reasonably interpreted as meaning that any cancellations occurring in the average values also occur in the covariances and variances. To the accuracy of this approximation, then, we appear to be using the data rather efficiently.

If, on the other hand, the true spectrum, $P(f)$, consists of a single sharp peak at $f = f_0$, we may use the approximation, derived in Section B.7, namely

$$\text{cov } \{P_i(f_1), P_j(f_2)\} \approx \left[\int_0^\infty P_{i1}(f) \cdot df \right] \cdot \left[\int_0^\infty P_{j2}(f) \cdot df \right]$$

$$\approx \text{ave } \{P_i(f_1)\} \cdot \text{ave } \{P_j(f_2)\},$$

a result which is not influenced by T'_n (so long as T'_n does not become large enough for $1/T'_n$ to become comparable with the width of the peak).

$$W_e = \frac{\left[\int_0^\infty P_{i1}(f) \cdot df\right]^2}{\int_0^\infty [P_{i1}(f)]^2 \cdot df}$$

is naturally called the *equivalent width* of $P_{i1}(f) = H_i(f; f_1) \cdot P(f)$.

The longer the record, and the wider the equivalent width, the more stable the estimate. (Increasing the width also of course makes the estimate refer to an average power density over a wider frequency interval.)

If, on the other hand, $P(f)$ consists of a sharp peak, then, by the concluding remarks of the preceding section

$$\frac{\text{var}\ \{P_i(f_1)\}}{[\text{ave}\ \{P_i(f_1)\}]^2} = 1.$$

The equivalent widths of some simple cases are as follows:

1. If $P_{i1}(f)$ is a rectangle of width W which does not cross $f = 0$, then $W_e = W$.

2. If $P_{i1}(f)$ is a triangle of base W which does not cross $f = 0$, vertex anywhere over the base, then $W_e = 0.75\ W$.

3. If $P_{i1}(f)$ is proportional to

$$\frac{\sin \dfrac{\omega + \omega_1}{W}}{\dfrac{\omega + \omega_1}{W}} + \frac{\sin \dfrac{\omega - \omega_1}{W}}{\dfrac{\omega - \omega_1}{W}}$$

i.e. has the shape of $H_0(f; f_1)$, where $W = W_{\text{main}} = 2W_{\text{side}}$ (these being the widths of main and side lobes, respectively), and if $f_1 \geqq 1/T_m$ then $W_e = 0.5\ W = 0.5\ W_{\text{main}} = W_{\text{side}}$.

4. If $P_{i1}(f)$ has the shape of $H_2(f; f_1)$, i.e. is proportional to a hanning

(0.25, 0.5, 0.25) window, and if $f_1 \geqq 1/T_m$, then $W_e = 0.67\,W_{main} = 2.67\,W_{side}$.

5. If $P_{i1}(f)$ has the shape of $H_3(f; f_1)$, i.e. is proportional to a hamming (0.23, 0.54, 0.23) window, and if $f_1 \geqq 1/T_m$, then $W_e = 0.63\,W_{main} = 2.52\,W_{side}$.

These cases are illustrated in Fig. 3, a single sketch sufficing for the last two. Note that W_e is close to $\frac{2}{3}W_{main}$ for practical windows, if $f_1 \geqq 1/T_m$.

For our standard window pairs, hanning or hamming, the width of the normal side lobes is $1/(2T_m)$ and, consequently, $W_e \sim 1.30/T_m$, if $f_1 \geqq 1/T_m$.

These last three equivalent widths decrease somewhat as f_1 becomes small, and the values given should be halved for $f_1 = 0$.

If $P(f)$ varies linearly across $H_i(f; f_1)$, then a calculation discussed in Section B.8 shows that W_e will tend to fall in the range from $1.15/T_m$ to $1.23/T_m$. (A rather peaked case gives $0.94/T_m$.) When we allow for the fact that we are likely to be concerned with processes which are not quite Gaussian, whose variances of estimate are consequently likely to be somewhat larger than for the Gaussian case, a change corresponding to the use of a decreased equivalent width in the formula, the choice

$$W_e \approx \frac{1}{T_m}$$

which introduces a small factor of safety (not more than 1.3) seems de-

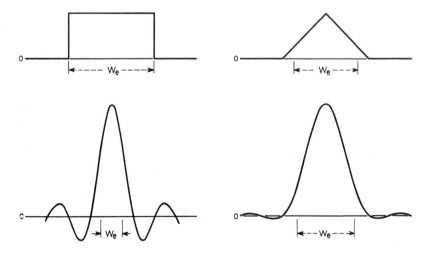

Fig. 3 — Equivalent widths of some spectral windows.

sirable for planning purposes. Consequently, we shall plan according to

$$\frac{\text{var } \{P_i(f_1)\}}{[\text{ave } \{P_i(f_1)\}]^2} \approx \frac{T_m}{T'_n}.$$

the noise being exactly Gaussian. Real noises (and especially real signals) need not be exactly Gaussian. Thus, even exact results in Gaussian theory would be approximations in practice. Second, the chief purposes of studying variability are first to choose, once for all, effective methods of analysis, and then, in each situation, to determine about how much data will be required for the desired or given accuracy. Again, approximate results will be adequate. Third, it would not be safe to use the advance estimates of variability as firm, guaranteed, measures of the stability of the actual computed results in a practical situation, since other sources of variability may well contribute to the deviation of a particular spectral density estimate from its long run value. (Non-constancy of total power level, even with distribution-over-frequency remaining constant, and failures of stationarity are two simple examples.) We must rely on observed changes from trial to trial as basically the safest measure of the lack of stability of our spectral density estimates.

Thus, the purposes of variability theory are well served if its results are approximate — deviations of actual variability from theoretical variability of ±5 per cent, ±10 per cent or even ±20 per cent will be quite satisfactory. Judged by this standard, the variability theory based on (i) the Gaussian assumption and (ii) treating the distribution of the spectral density estimates as if they followed so-called "chi-square" distributions, as we shall do in the next section, will usually be very satisfactory.

9. CHI-SQUARE — EQUIVALENT DEGREES OF FREEDOM

If y_1, y_2, $\cdots$, y_k are independently distributed according to a standard normal distribution, that is, according to a Gaussian distribution with average zero and unit variance (and, consequently, unit standard deviation), then

$$\chi_k^2 = y_1^2 + y_2^2 + \cdots + y_k^2,$$

which is obviously positive, follows, by definition, a chi-square distribution with k degrees of freedom. The coefficient of variation of χ_k^2, the ratio of RMS deviation to average value, is $(2/k)^{1/2}$, so that, as k increases, χ_k^2 becomes *relatively* less variable. This statement also applies to any multiple of χ_k^2.

A convenient description of the stability of any positive or nearly-positive estimate is its *equivalent number of degrees of freedom*, the number of degrees of freedom of that χ_k^2 some multiple of which it resembles (in average and variance unless otherwise specified). We can find such a k from

$$k = \frac{2(\text{average})^2}{\text{variance}} = \frac{2}{(\text{coefficient of variation})^2}.$$

Interpretation is aided by Tables I and II. These tables are possible because the distribution of the ratio of any multiple of χ_k^2 to the average value (of that multiple) depends only on k. Thus, if $k = 4$, individual

TABLE I

Distribution of quantities which are distributed as fixed multiple of chi-square. Ratios of individual value to its average value exceeded with given probabilities.

Degrees of freedom	Exceeded by 90% of all values	Exceeded by 50 % of all values	Exceeded by 10% of all values
1	0.016	0.46	2.71
2	0.10	0.70	2.30
3	0.19	0.79	2.08
4	0.26	0.84	1.94
5	0.32	0.87	1.85
10	0.49	0.93	1.60
20	0.62	0.96	1.42
30	0.69	0.98	1.34
40	0.73	0.98	1.30
50	0.75	0.99	1.26
100	0.82	0.99	1.18
200	0.873	1.00	1.139
500	0.920	1.00	1.081
1000	0.943	1.00	1.057

Examples: (1) If the long run average is 10 volts²/cps, then among estimates with 10 degrees of freedom, 10 per cent would fall below 4.9 volts²/cps, and 50 per cent would fall above 9.3 volts²/cps.
(2) If a single observed estimate, with 5 degrees of freedom, is observed to be 2 volts²/cps, then we have 80 per cent confidence that the true long-run value lies between 2/1.85 = 1.08 volts²/cps and 2/0.32 = 6.25 volts²/cps.

TABLE II — BEHAVIOR OF χ_k^2 ON DECIBEL SCALE

Fraction of distribution	Spread* of interval† in db‡	k required for interval† of spread		

(spread is the difference between the upper boundary expressed in db, and the lower boundary expressed in db.)

† All intervals are symmetric in the probability sense, half of the non-included probability falling above and half below the interval.

‡ Since we are dealing with measures of variance, analogous to power, 10 db = (factor of 10), and (number of db) = (10 $\log_{10}$ ratio of variances).

values of any particular multiple of χ_4^2 will, in the long run, fall below 0.26 times their average value in 10 per cent of all cases (will be 5.8 db or more below average in 10 per cent of all cases). Similarly, individual values will, in the long run, fall below 0.84 times their average value (be 0.7 db or more below average) in 50 per cent of all cases, and in 90 per cent of all cases will fall below 1.94 times their average value (be 2.9 db or less above average). Thus, in the long run, 80 per cent of all values will fall in an interval of spread $(2.9) - (-5.8) = 8.7$ db.

Thus, for example, to obtain 4 chances in 5 that a single observed value will lie within ±30 per cent of the true value we require (see Table I) about 40 degrees of freedom, while to obtain 4 chances in 5 that a single observed value will lie in a prescribable interval of length 5 db, we require (see Table II) at least 11 degrees of freedom.

The results of the preceding section indicate that, for an estimate of smoothed spectral density, when $P(f)$ is smooth, the number of degrees of freedom is given by

$$k = 2T'_n W_e = \frac{W_e}{\Delta f},$$

where the latter form expresses the number of degrees of freedom as the number of *elementary frequency bands*, each of width

$$\Delta f = \frac{1}{2T'_n},$$

contained in the equivalent width W_e.

For design purposes, the relation of the last section (including the small safety factor) indicates that

$$k = \frac{2}{[\text{var } \{P_i(f_1)\}]/[\text{ave } \{P_i(f_1)\}]^2} \approx \frac{2T'_n}{T_m}$$

when $P(f)$ varies slowly. (This will usually be the case if (i) $k > 3$ or 4, say, and (ii) $P_{i1}(f)$ is a moderately smooth single hump. For, under these circumstances, $P_{i1}(f)$ will not change rapidly in a frequency interval $1/T'_n$ and the same property can then be inferred for $P(f)$ itself.)

When, on the other hand, $P(f)$ consists of a single sharp peak, we find, using the last result of Section 7, that $k \approx 2$, so long as $1/T'_n$ is not small enough to be comparable with the width of the peak. At first glance, this result may appear a little surprising, but when we notice that a single spectral line corresponds either (a) to frequency $+f_0$ *and* to frequency $-f_0$, or (b) to cos $\omega_0 t$ *and* to sin $\omega_0 t$, or (c) to amplitude *and* to phase, it appears quite natural that a sharp line carries *two* degrees of freedom and not merely one.

We may summarize the semi-quantitative study of the stability of estimates of the smoothed power spectrum as follows:

(a) It is not necessary to judge stability with very high accuracy.

(b) It is convenient to measure stability by analogy with the number of degrees of freedom associated with a multiple of a chi-square variate.

(c) The equivalent number of degrees of freedom can be regarded as the number of elementary bands of width Δf in the equivalent width W_e of the *filtered* spectrum

$$2P_{i1}(f) = H_i(f; f_1) \cdot 2P(f) \qquad (f \geqq 0)$$

if the result is not too small (say > 3 or 4) and $P_{i1}(f)$ is moderately smooth.

(d) If the filtered spectrum approaches a single sharp peak, the equivalent number of degrees of freedom for the corresponding estimate approaches two.

In interpreting the concept of equivalent number of degrees of freedom, it may be helpful to imagine the continuous density of the *filtered* spectrum replaced by a discrete set of ordinates, one per elementary frequency band. If these ordinates are p_0, p_1, p_2, $\cdots$, the natural approximation to the number of degrees of freedom is

$$k = \frac{(p_0 + p_1 + p_2 + \cdots)^2}{p_0{}^2 + p_1{}^2 + p_2{}^2 + \cdots}$$

as illustrated in Fig. 4. This approximation will usually be satisfactory

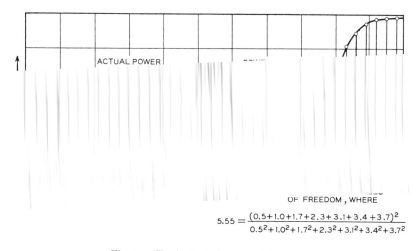

Fig. 4 — Equivalent degrees of freedom.

as long as the effect on k of moving each ordinate around within its elementary frequency band can be neglected. (In more extreme cases, an approximation based on two ordinates per pair of elementary frequency bands is more precise.)

10. DIRECT ANALOG COMPUTATION — GRADED DATA WINDOWS

We have been dealing thus far with continuous time, and the communications engineer will naturally ask, "Why introduce autocovariance functions and all that, why not measure the spectrum by filtering, rectifying, and smoothing?". The only fair answer is "By all means, do so if you can obtain, and maintain, the necessary accuracy economically!" Let us apply our results to such a measurement technique.

Let $X(t)$ be the noise or signal whose power spectrum $P(f)$ we wish to study. Let us pass it through a filter of transfer function $Y(f)$, and designate the result by $X_{out}(t)$. Its power spectrum, $P_{out}(f)$, will be given by

$$P_{out}(f) = |Y(f)|^2 \cdot P(f)$$

and if a section of $X_{out}(t)$ of length T_n is applied to an ideal quadratic rectifier and smoothed by a smoothing network of infinite time constant, the result will be

$$\int_0^{T_n} [X_{out}(t)]^2 \, dt.$$

The average value of this result divided by T_n is

$$\int_0^\infty 2P_{out}(f)\, df,$$

and the number of equivalent degrees of freedom is the number of elementary frequency bands, of bandwidth $1/(2T_n)$, contained by the equivalent width of $\mid Y(f) \mid^2 \cdot P(f)$. This last function is of the form

(power transmission function)(original power spectrum)

just as before. We see that the ideal process of filtering, rectifying, and smoothing the actual input has produced the same accuracy as the ideal process of calculating, modifying, and transforming the apparent autocovariance, provided that $\mid Y(f) \mid^2 = H_i(f; f_1)$ for a suitable choice of T_m, f, and f_1. This is what we ought to have expected, since we believe that either method extracts nearly all the information about the spectrum which the data provides.

A few practical considerations deserve mention. They center around the actual switching situations which can arise, especially when we have only a finite sample of the original noise. In Fig. 5, the watt-second meter includes quadratic rectification and integration functions which we think of as ideal. (It may be very important to allow for the fact that the "ground" position of switch A is not quite at the same potential as the zero of the input noise, but we shall neglect this effect for the moment.)

Some four sorts of operation can arise according to the times at which switch B is operated. The watt-second meter may be connected either at the beginning of the running period T or after some interval of time (to allow initial transients to become negligible), and may be disconnected either at the end of the running period T or after some interval of time (to allow the meter to reach a maximum). These four modes of operation are illustrated in Fig. 6.

In Mode I, providing the initial waiting period is long enough to allow transients to become negligible, the filter output is essentially stationary, and the earlier discussion in this section applies.

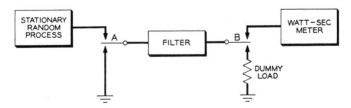

Fig. 5 — Schematic analog circuit.

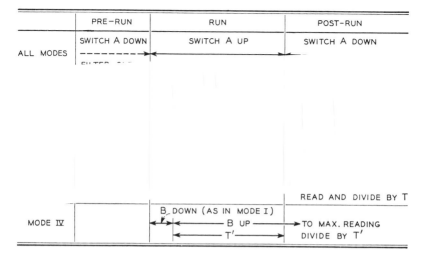

Fig. 6 — Time histories of operation for different modes.

In Mode II, all of the energy output is recorded on the meter, but the reading is divided only by the length of the input data. This mode is amenable to exact and complete analysis which is given in some detail in Section B.10. The results differ from those of Mode I in that the transform of the boxcar function of length T (running period) is convolved twice into the spectral window. (Convolution is defined and discussed in Section A.3.) If T is not large, the effects may be somewhat uncomfortable in that the spectral window becomes wider and more ragged.

Mode III, discussed briefly in Section B.10, differs from Mode II by an additional convolution whose effect again disappears as $T \to \infty$.

Mode IV resembles Mode I in that the noise input is passed through the filter until transient effects have become negligible, when the meter is switched on at the filter output. It differs from Mode I in that the meter is read *after* a final waiting period. This seems to offer no advantages over Mode I, and will not be discussed further.

The contrast between Mode I and Mode II is another example of what should now be becoming familiar. Mode I has no additional convolution in the spectral window. Mode II provides data economy by making it possible to integrate over the whole length of the available record. We should really like both advantages.

We can, indeed, obtain most of both advantages, but only by replacing the sharp edges of the switched data window by the smoothed outlines of a graded data window. In other words, we need to introduce

$$X_{in}(t) = B(t) \cdot X(t)$$

at the input of the filter, where $B(t)$ vanishes except for $0 < t < T$, and is smooth enough to have its Fourier transform $J(f)$ concentrated near $f = 0$. Details are discussed in Section B.10.

Difficulties arising from the fact that the zero of the $X(t)$ input might not be at ground are shown in Section B.10 to behave similarly to those arising from switching transients, namely, no effect in Mode I, possibly uncomfortable in Modes II and III, usually negligible when a well-chosen graded data window is used.

Another device is sometimes used to make maximum use of a finite noise record. The record is merely closed into a continuous loop, and the rectifier-smoother output averaged. It is shown in Section B.10 that here, too, we must use a graded data window $B(t)$.

11. DISTORTION, NOISE, HETERODYNE FILTERING AND PREWHITENING

Another group of very important practical considerations center around the spectrum of the "signal" as it is handled (either instantaneously, or in recorded form). We have spoken of "filtering, rectifying and smoothing" and have treated all these steps as ideal. No attention has been given to the equally vital "gathering" and "transmission and recording" steps. Tacitly, they too have been treated as ideal. Realistically, we must expect a certain amount of distortion (nonlinearity, intermodulation, etc.) and the addition of a certain amount of background noise in all three of the first steps: gathering, transmission and recording, filtration. It often proves to be most important to lessen the ill effects of such distortion and noise addition.

In a perfect system, and with a fixed spectral window, the fluctuations of an estimate are proportional to its average value. If we have a fixed uniform noise level, it will do the least additional damage if all the average values of the estimates are of about the same size, for then no low estimate can "disappear" into the noise.

Intermodulation distortion will have the greatest effect on the signal being transmitted when two strong frequencies combine to produce a modulation product whose frequency falls in a very weak region of the spectrum, for it is in such situations that the fractional distortion of the spectrum reaches its maximum. To minimize possible effects of intermodulation distortion it is again desirable to transmit, record and generally handle signals with a roughly flat spectrum.

To these noise and intermodulation considerations another sort of consideration may be added. Many frequency analyzers use a heterodyne system, bringing the frequency band to be studied to a fixed filter, rather than tuning a filter across a wide frequency band. The power transfer function of the combination of heterodyne modulator and fixed filter, referred to input frequency, will depend only on Δf, the d... ...

..., ,,,), ...practically equal to $Q_i(f - f_1)$, in order to attain this as f leaves f_1. We may thus be forced to compromise properties of $Q_i(\Delta f)$ useful near other frequencies. The simplest way to avoid such problems is to arrange for the $P(f)$ of the "signal" analyzed to be fairly constant, or at most slowly varying.

Thus, for a variety of reasons, we can often gain by introducing "compensation" or "preemphasis" to make more nearly constant the spectrum of the "signal" actually transmitted or recorded, and analyzed. Since the ideal would be to bring the spectrum close to that of white noise, it is natural to refer to this process as *prewhitening*. Such flattening of the spectrum need not be precise, or even closely approximate. We need only to make the rate of change of $P(f)$ with frequency relatively small.

Because of advantages related to the noise and intermodulation distortion introduced in various steps of the sequence, it will be best, other considerations aside, to carry out such prewhitening at as early a point in the measurement-analysis sequence as possible. Sometimes this can even be done in the pick-up or sensing element.

This whole philosophy of prewhitening, which appears quite natural to the communication engineer familiar with preemphasis and other techniques for increased information transfer within a given frequency interval, comes as a great change to the instrumentation engineer, whose clients ordinarily require "faithful" reproduction of an input at the output, by which they mean phase shifts nearly linear with frequency, and a nearly constant amplitude response up to some high frequency. It will be rare indeed, in practical spectrum analysis, that the ideal response for the initial transducer and amplifier will be flat. Instead it should have a characteristic contributing to prewhitening. This characteristic will, of course, have to be measured separately and the corresponding adjustments to the estimates of the spectral density will have to be made so that these estimates, instead of applying to the "signal" actually analyzed, apply to the original "input signal", but such labor will often be many times repaid.

One further consideration about frequency responses in measurement now enters naturally. In almost every power spectrum problem there is an upper frequency beyond which there is no appreciable interest. In most components used in measurement, transmission, recording, etc., the noise level, and often the level of intermodulation distortion, is roughly a fixed fraction of the peak useful level. If substantial power is present at frequencies so high as to be uninteresting, then the need to keep *total* power below the peak useful level forces us to handle the interesting frequencies at a power level below that which could otherwise be used. The ratio of noise and intermodulation distortion to interesting signal is thus raised — the quality of the analysis and its results degraded. The appropriate remedy is to filter out the uninteresting high frequencies at as early a stage as possible. This is a further reason why a carefully tailored frequency response is an important part of a power spectrum measuring process.

Together with the need for *adequately wide filters* (we can of course use narrower filters when we are prepared to average over homogeneous records of sufficiently long total duration) to provide enough equivalent degrees of freedom, and hence enough stability for the estimates, this tailoring of frequency response is often the crucial part of a power spectrum measuring program. Indeed, there may sometimes be no reasonable way to measure power spectra with an ill-tailored frequency response, even if this response be "flat".

EQUALLY SPACED RECORDS

We come now to treat a modified situation of great practical importance, where the observations are used for analysis only at equally spaced intervals of time — not as a continuous time record. Two new and important features enter: there is aliasing of frequencies, and practical analysis will involve digital rather than analog computation. In general, however, the situation is surprisingly similar to the case of a continuous record, with limitations on data-gathering effort still forcing us to compromise resolution and stability. Advantages of convenient calculation and noise reduction still lead us to prewhitening. Filtering of equi-spaced data must involve transversal filters (see Glossary of Terms for definition) whose transmission properties (in frequency) exhibit a periodic symmetry. This exerts additional pressure toward prewhitening.

Questions regarding computational techniques arise anew because of the nature of digital computation. These include means for reducing the effects of a displaced (perhaps drifting) zero, smoothing by groups to economize arithmetical operations, ~~ *1 ~ ~ ~

$$| \tau | = 0, \Delta t, 2\Delta t, \cdots, n\Delta t.$$

Now, the equations

$$C(\tau) = \int_0^\infty 2P_A(f) \cdot \cos 2\pi f \tau \cdot df,$$

$$| \tau | = q\Delta t, \qquad q = 0, 1, \cdots, n,$$

if soluble at all, can always be satisfied by a $P_A(f)$ which vanishes for $f > f_N = 1/(2\Delta t)$, although the power spectrum $P(f)$ of the original process (for which the $C(\tau)$ was defined) might actually cover a much wider frequency range. (We shall reserve the notation $P_A(f)$ for such a function, vanishing for $|f| > f_N$.) While frequencies between $f = 0$ and $f = f_N$ are clearly distinct from one another, we face a problem of aliasing, since frequencies above f_N usually contribute some power. Each frequency, no matter how high, is indistinguishable from one in the band from 0 to f_N.

The essential, unavoidable nature of this problem is made clear by Fig. 7 which illustrates how equally spaced time samples from any

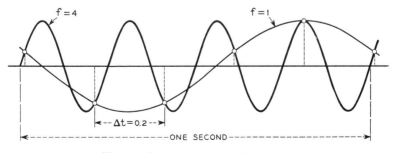

Fig. 7 — Sampling of sinusoidal waves.

cosine wave *could* have come from each of many other cosine waves. (The familiar stroboscope uses a particular expression of this fact in apparently "slowing down" rapidly rotating or oscillating machinery.) The logical position about $P_A(f)$ depends very much on whether $X(t)$ is thought of as having any real existence for $|t| \neq q\Delta t$.

If $X(t)$ really exists for continuous t, although we have (i) failed to observe or record it, or (ii) failed to "read" the record, or (iii) decided to neglect the available values, then there is a well-defined $P(f)$ corresponding to the process from which each $X(t)$ is a sample, and we must be very careful about the relation between $P(f)$, which is our true concern, and $P_A(f)$, which is clearly all we can strive to estimate directly from the data. It can be shown (see Section B.12) that, in the form appropriate for a one-sided spectrum, if we set

$$2P_a(f) = 2P(f) + 2P(2f_N - f) + 2P(2f_N + f)$$
$$+ 2P(4f_N - f) + 2P(4f_N + f) + \cdots$$

then we may take

$$P_A(f) = \begin{cases} P_a(f), & 0 \leq |f| \leq f_N \\ 0, & \text{otherwise} \end{cases}$$

where $f_N = 1/(2\Delta t)$ is the *folding* (or Nyquist) frequency. We naturally call the frequencies f, $2f_N - f$, $2f_N + f$, $4f_N - f$, $4f_N + f$, and so on, *aliases* of one another, f being the *principal alias*. The *aliased spectrum* $P_a(f)$ is the result of *aliasing* $P(f)$. The *principal part of the aliased spectrum* $P_A(f)$ is the part of $P_a(f)$ which corresponds to principal aliases, positive and negative.

(If $X(t)$ has no natural existence for t's which are not integral multiples of Δt, then $P(f)$ is not uniquely defined, and we are at liberty to choose any normalization we desire. In particular, we may decide to limit $P(f)$ to the interval $|f| \leq 1/(2\Delta t)$, in which case we will be enforcing $P(f) = P_A(f)$ without any trace of aliasing. We mention this case for logical completeness, but remark that it seems to occur infrequently in practice, whatever the field.)

If the Gaussian noise we are considering has a power spectrum $P(f)$ which extends outside $|f| \leq 1/(2\Delta t)$, then the Gaussian noise with spectrum $P_A(f)$ is not the same for continuous time. However, if we consider these two noises only for equi-spaced times

$$t = 0, \Delta t, 2\Delta t, \cdots$$

they are identical. For all first moments vanish and all second moments coincide, which implies coincidence of the joint distributions of any finite set from $\cdots, X_{-q}, \cdots, X_{-1}, X_0, X_1, \cdots, X_q, \cdots$, and this is our definition of the coincidence of two noises. (If a result concerning such equally spaced values can be established, ...

une effective length, then, since

$$\frac{f_N}{\text{elementary frequency bandwidth}} = \frac{\dfrac{1}{2\Delta t}}{\dfrac{1}{2T'_n}} = n'$$

there are n' elementary frequency bands between 0 and f_N. As a statistician would have anticipated, we gain one elementary frequency band — one degree of freedom — for each added observation.

It is perhaps natural to base a hope on this result — a hope that taking data more frequently over the same time interval would gain us many degrees of freedom and reduce our difficulties with variability. However, this is not the case (as the expression for the width of an elementary frequency band $1/(2T'_n)$ should have warned us). Taking observations twice as frequently yields twice as many elementary frequency bands, but also doubles the folding frequency f_N and, thus, doubles the frequency interval occupied by principal aliases. The density of elementary frequency bands is not increased one iota. (Clearly, iota was the Greek word for bit!).

It is usual for aliasing to be present and to be of actual or potential importance. This is an inescapable consequence of data taken or read at uniform intervals. (It is not infrequently suggested that there should be a workable scheme of taking discrete data in some definite, but not uniformly spaced pattern, and estimating the power spectrum without aliasing. No such scheme seems so far to have been developed).

13. TRANSFORMATION AND WINDOWS.

Given uniformly spaced values of $X(t)$ — values which we shall now designate $X_0, X_1, \cdots, X_n$, as well as $X(0), X(\Delta t), \cdots, X(n\Delta t)$ — we expect to calculate "sample autocovariances", modify them, and then Fourier transform the results. There is no possibility of calculating autocovariances for lags other than 0, Δt, $\cdots$, $n\Delta t$, and so we may as well write C_0, C_1, $\cdots$, C_m in place of $C_i(0)$, $C_i(\Delta t)$, $\cdots$, $C_i(m\Delta t)$. If we Fourier transform these $m + 1$ numbers, as obtained or modified, we might obtain a smoothed spectral estimate for any frequency between 0 and $f_N = 1/(2\Delta t)$ that we may wish. It is not surprising, however, that we lose no information (and little explication) if we calculate only $m + 1$ such estimates (one for each C_r). Nor is it surprising that we regularly take these estimates equally spaced over $0 \leq f \leq f_N$, and hence at intervals of $f_N/m = 1/(2m\Delta t)$. As a consequence we have to deal with finite Fourier (cosine) series transformation (classical harmonic analysis) rather than with infinite Fourier integral transformation, but the correspondence between multiplication and convolution persists.

The question of modification also requires discussion. In the continuous case we Fourier transformed

$$C_i(\tau) = D_i(\tau) \cdot C_{00}(\tau) = D_i(\tau) \cdot C_0(\tau)$$

where $C_0(\tau)$ coincided with $C_{00}(\tau)$ wherever the latter was defined, and is zero otherwise (cp. Section B.5). The result was, consequently (e.g. see Section A.3), the convolution of the Fourier transforms of $D_i(\tau)$ and $C_0(\tau)$. So long as time was continuous and computation was presumably by analog devices, there was a real advantage to modification before transformation. Now that time is discrete and computation presumably digital, the advantage is transferred to first transforming and then convolving. Indeed, because the $D_i(\tau)$, for $i > 1$, are finite sums of cosines, so that their transforms are simply sums of spikes (Dirac delta-functions) at the appropriate spacing, convolution means only smoothing with weights

$$0.25, 0.5, 0.25 \qquad (i = 2, \text{hanning})$$

$$0.23, 0.54, 0.23 \qquad (i = 3, \text{hamming})$$

and is very simply carried out.

In discussing this program, we gain some generality by using $m + 1$ lags separated by $\Delta\tau = h\Delta t$ for an integer $h > 0$, while our results are no more complicated than if we were to confine ourselves to $h \equiv 1$, which is the practical case. Thus, we first compute the *mean lagged products*

$$C_r = \frac{1}{n - rh} \sum_{q=0}^{q=n-rh} X_q \cdot X_{q+rh}$$

for $r = 0, 1, 2, \cdots, m$, where $mh < n$. Note that C_r is heuristically as close as we can come to the apparent autocovariance $C_{00}(\pm r\Delta\tau)$ with the available (equi-spaced) data. Note further that, so far as functions of the C_r are concerned, our effective folding frequency is

$$V_r = \Delta\tau \cdot \left[C_0 + 2 \sum_{q=1}^{m-1} C_q \cdot \cos \frac{qr\pi}{m} + C_m \cdot \cos r\pi \right].$$

(We may regard this as arising from replacing $C_0(\tau)$ in the expression for $P_0(f)$ as its Fourier integral transform by a finite sequence of spikes (Dirac delta functions) of intensities (areas) proportional to the corresponding values of $C_0(\tau)$.) If we put

$$P_{0A}\left(\frac{r}{2m \cdot \Delta\tau} \right) = V_r$$

then it is shown in Section B.13 that

$$\text{ave } \{P_{0A}(f)\} = \int_{-\infty}^{\infty} Q_0(f - f'; \Delta\tau) \cdot P(f') \cdot df'$$

where

$$Q_0(f; \Delta\tau) = \Delta\tau \cdot \cot \frac{\omega\Delta\tau}{2} \cdot \sin m\omega\Delta\tau.$$

In terms of $Q_0(f)$, which is treated in Section B.5, we have

$$Q_0(f; \Delta\tau) = \sum_{q=-\infty}^{\infty} Q_0\left(f - \frac{q}{\Delta\tau} \right) = Q_{0A}(f).$$

Just as the average value of $P_0(f)$ in the continuous case is the corresponding value of $Q_0(f) * P(f)$, so here the average value of $P_{0A}(f)$ is the corresponding value of $Q_0(f; \Delta\tau) * P(f)$. Thus, we may consider $P_{0A}(f)$ as estimating the result of "smoothing" $P(f)$ with a window $Q_0(f; \Delta\tau)$ which has repeated major (and concomitant minor) lobes at intervals of $2f_N^* = (\Delta\tau)^{-1}$. This is not the most convenient way to consider matters, and in Section B.13 it is shown that there are two equivalent forms for ave $\{P_{0A}(f)\}$ and, correspondingly, two other, equally appropriate, ways to consider the situation.

These arise from the three-fold identity

$$Q_{0A}(f) * P(f) \equiv Q_0(f) * P_a(f) \equiv Q_{0A}(f) * P_A(f),$$

any member of which represents the average value of $P_{0A}(f)$. Thus, we can also consider $P_{0A}(f)$: (i) as estimating the result of smoothing the infinite, periodic aliased spectrum $P_a(f)$ with the same window as for the continuous case, or (ii) as estimating the result of smoothing the principal part of the aliased spectrum $P_A(f)$ with the aliased window $Q_{0A}(f)$. The latter choice is usually the most helpful of the three possibilities, and is the one we shall adopt.

All this has been discussed for the immediate results of transforming unmodified C_r's. This is only the case $i = 0$ of the identity

$$Q_{iA}(f) * P(f) \equiv Q_i(f) * P_a(f) \equiv Q_{iA}(f) * P_A(f)$$

which holds in general. We should thus usually be concerned with $Q_{iA}(f)$ and with $P_A(f)$.

The case $i = 2$ (hanning) corresponds to the following smoothing after transformation:

$$U_0 = 0.5\ V_0 + 0.5\ V_1,$$

$$U_r = 0.25\ V_{r-1} + 0.5\ V_r + 0.25\ V_{r+1}, \qquad 1 \leqq r \leqq m - 1,$$

$$U_m = 0.5\ V_{m-1} + 0.5\ V_m,$$

for which $Q_{2A}(f)$ has the form shown in Fig. 8. The curve is for $m = 12$, and the circles are for $m = \infty$, which corresponds exactly to the continuous case. Clearly, for usual values of m, the modification in the lobes due to aliasing is almost surely unimportant.

The frequency separation between adjacent estimates is

$$\frac{1}{2T_m} = \frac{1}{2m\Delta\tau},$$

but the equivalent width of the windows (for $1 \leqq r \leqq m - 1$) is about

$$\frac{1.30}{T_m} = \frac{1.30}{m\Delta\tau},$$

just as for the continuous case (see Section 8). For most purposes we may again take the bandwidth corresponding to each estimate as $1/T_m$,

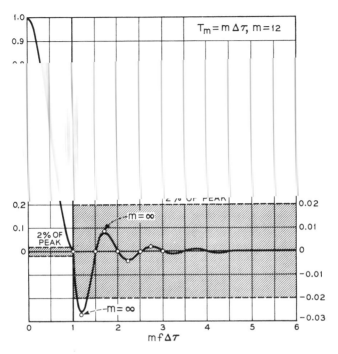

Fig. 8 — Aliased spectral window Q_{2A} for $m = 12$.

so that m satisfies

$$\frac{1}{m} = \text{(bandwidth of estimates)} \cdot (\Delta\tau).$$

If we had neither modified before Fourier transformation, nor smoothed after transformation, we should have faced the uncomfortable minor lobes of $Q_{0A}(f)$ shown in Fig. 9 for $m = 12$ (with circles for $m = \infty$). Generally speaking, all we learned about desirable lag windows for the continuous case carries over with minor modifications, at most. The only serious effect of going to uniformly spaced values is the aliasing (and this may be very serious indeed).

It is well worth noting that the possible spectral windows $Q_{iA}(f)$ are now restricted to be finite Fourier series in $\cos \omega\Delta\tau$, $\cos 2\omega\Delta\tau$, $\cdots$, $\cos m\omega\Delta\tau$, or equivalently, to be polynomials in $\cos \omega\Delta\tau$ of degree m at most.

14. VARIABILITY AND COVARIABILITY

We now extend all our other notation: $H_i(f; f_1)$, $P_i(f_1)$, etc. to corresponding $H_{iA}(f; f_1)$, $P_{iA}(f_1)$, etc. for the uniformly spaced case as

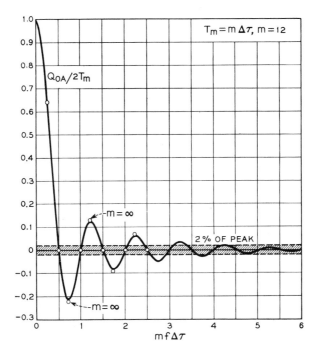

Fig. 9 — Aliased spectral window Q_{0A} for $m = 12$.

specified in Sections B.13 and B.14. It is shown in the latter section that we now have

$$\text{cov} \{2P_{iA}(f_1), 2P_{jA}(f_2)\} = \int_0^\infty H_{iA}(f; f_1) \cdot H_{jA}(f; f_2) \cdot 2\Gamma_{\Delta t}(f) \cdot df$$

where

$$\Gamma_{\Delta t}(f) \approx 4 \int_{-\infty}^\infty P_A(f' + f) \cdot P_A(f' - f) \cdot \left(\frac{\sin \omega' T_n'}{\omega' T_n'}\right)^2 \cdot \left(\frac{\sin \omega' \Delta t}{\omega' \Delta t}\right)^{-2} df',$$

$(\omega' = 2\pi f')$, with a very slightly different determination of T_n' than before. The only essential change has been the introduction of a new factor, corresponding to aliasing,

$$\left(\frac{\sin \omega' \Delta t}{\omega' \Delta t}\right)^{-2}$$

into the integrand of the power-variance spectrum $\Gamma_{\Delta t}(f)$. For usable values of n, this factor will vary much more slowly than

$$\left(\frac{\sin \omega' T_n'}{\omega' T_n'}\right)^2$$

and can usually be treated as sensibly equal to unity. All the approximate analysis of covariability and variability given for the continuous case now goes through without essential change.

at equal intervals, either as a supplement to, or as a partial replacement for, early prewhitening.

The average value of a power density estimate $P_{iA}(f_1)$ is

$$\text{ave } \{P_{iA}(f_1)\} = \int_0^\infty P_{iA1}(f) \cdot df,$$

where

$$P_{iA1}(f) = H_{iA}(f; f_1) \cdot P_A(f).$$

We want this quantity to tell us about the values of $P(f)$ for f near f_1. To do this we must: (i) reduce variability, (ii) ensure that $P_A(f)$ resembles $P(f)$ sufficiently, and (iii) concentrate $P_{iA1}(f)$ near $f = f_1$. We must be concerned with: (i) adequately broad windows, (ii) sufficiently weak aliasing, and (iii) enough sharpness in the effective filter. This sharpness can be obtained in a combination of ways.

Note that we asked for $P_{iA1}(f)$, which measures the net contribution to the average value, to be localized. We did not merely ask that $H_{iA}(f; f_1)$ should be localized. For, if

$$P_A(f_2) >>> P_A(f_1),$$

although

$$H_{iA}(f_2; f_1) << H_{iA}(f_1; f_1),$$

it is still possible for

$$P_{iA1}(f_2) = H_{iA}(f_2; f_1) \cdot P_A(f_2)$$

to outweigh

$$P_{iA1}(f_1) = H_{iA}(f_1; f_1) \cdot P_A(f_1),$$

so that our estimate tells us more about $P(f)$ near $f = f_2$ than it does about $P(f)$ near $f = f_1$. To avoid such unfortunate situations either we must choose our window pair in a very particular manner (so as to make $H_{iA}(f_2 ; f_1)$ exceptionally small) or we must avoid $P_A(f_2) >>> P_A(f_1)$. Both courses are possible and sometimes necessary. Usually, the second course is simpler.

Following the second course is simple in principle. Given actual values X_q, we apply a selected linear procedure to obtain new values $\tilde{X}_q$ and analyze these. The aliased spectrum $\tilde{P}_A(f)$ of the $\tilde{X}_q$ differs from the aliased spectrum $P_A(f)$ of the X_q by a known multiplicative function of frequency. (See Section B.15 for details.) Thus, (i) we may convert estimates of $\tilde{P}_A(f)$ into estimates of $P_A(f)$, and (ii) we may choose the linear procedure to make the aliased spectrum $\tilde{P}_A(f)$ of the $\tilde{X}_q$ reasonably flat.

The simplest linear procedures are probably the formation of moving linear combinations and the construction of autoregressive series. A simple example of a moving linear combination is

$$\tilde{X}_q = X_q - \alpha X_{q-1} - \beta X_{q-2} - \gamma X_{q-3}$$

for which the relation between the spectra is

$$\frac{\tilde{P}_A(f)}{P_A(f)} = \frac{\tilde{P}(f)}{P(f)} = |\, 1 - \alpha e^{-i\omega \Delta t} - \beta e^{-i2\omega \Delta t} - \gamma e^{-i3\omega \Delta t}\,|^2$$

$$= \text{a cubic in } \cos \omega \Delta t.$$

A suitable moving linear combination will generate any desired non-negative polynomial in $\cos \omega \Delta t$.

A simple example of an autoregressive combination is

$$\tilde{X}_q = X_q + \lambda \tilde{X}_{q-1} + \mu \tilde{X}_{q-2} + \nu \tilde{X}_{q-3}$$

for which the relation (reciprocal to that just considered) between the spectra is

$$\frac{\tilde{P}_A(f)}{P_A(f)} = \frac{\tilde{P}(f)}{P(f)} = \{|\, 1 - \lambda e^{-i\omega \Delta t} - \mu e^{-i2\omega \Delta t} - \nu e^{-i3\omega \Delta t}\,|^2\}^{-1}$$

$$= (\text{a cubic in } \cos \omega \Delta t)^{-1}.$$

A suitable autoregressive combination will, when indefinitely continued, generate the reciprocal of any desired non-negative polynomial in $\cos \omega \Delta t$.

By combining a suitable moving linear combination with suitable autoregression, as for instance in

$$\tilde{X}_q = X_q - \alpha X_{q-1} + \lambda \tilde{X}_{q-1} ,$$

which may also be written

$$\tilde{X}_q - \lambda \tilde{X}_{q-1} = X_q - \alpha X_{q-1},$$

for which

$\tilde{P}_{\cdot}(f)$ ~

_ .ιυιιrary non-negative

...υιι to multiply by any (simple) non-negative rational function of cos $\omega \Delta t$ is very substantial freedom. If we have a rough idea (see Section 18) of the behavior of $P_A(f)$, and if this behaviour is moderately smooth, though perhaps quite steep in places, we can usually do a very good job of flattening the spectrum by prewhitening after obtaining discrete (digital) values. Unless still bothered with steep slopes, we will usually then find that hanning, with its (0.25, 0.50, 0.25) weights and lower outer lobes is slightly preferable to hamming, with its (0.23, 0.54, 0.23) weights and reduced first minor lobes.

The main purpose of prewhitening *after* data has been obtained in digital form at equally spaced intervals is to avoid difficulty with the minor lobes of our spectral windows. We may regard the whole process of prewhitening, analysis with standard spectral windows, and, finally, compensation of estimate, as a means of constructing a set of specially shaped spectral windows, one for each center frequency, specially adapted to the data we are processing. This point of view is illustrated in Fig. 10. The uppermost curve shows the power transfer function of a hypothetical prewhitening filter, one which enhances mid-frequencies in comparison with those lower and higher. The next line shows two standard spectral windows, with symmetrical side lobes. The third line shows the effective spectral windows when prewhitening is followed by standard analysis, as given by the product of prewhitening power transfer function and spectral window. In either case, the side lobe toward mid-frequencies is higher than the corresponding side lobe on the opposite side, which is lower than for the standard. The lowest curve shows alternative spectra for time series which might reasonably be processed by the combination of prewhitening and standard analysis shown (since the prewhitened spectra would change only slowly). In every case, the side lobes of the special spectral windows are automatically so related to these spectra, as to balance and reduce the amount of leakage through them, as given by the product of special spectral window side lobe and original spectral density.

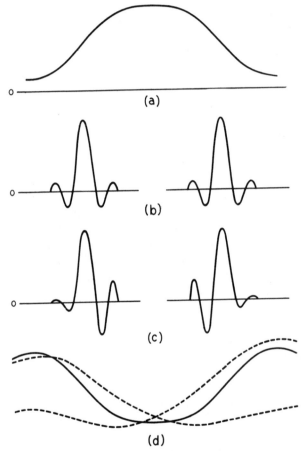

Fig. 10 — Illustration of prewhitening; (a) prewhitening power transfer function, (b) standard spectral windows, (c) effective spectral windows, and (d) typical input spectra to which (a) might be applied.

Easing of requirements for accuracy (number of significant figures, etc.) during computation are ordinarily quite secondary, though pleasant, advantages of prewhitening during digital calculation.

16. REJECTION FILTERING AND SEPARATION

If the difficulties in handling $P(f)$ are due, wholly or in part, to one or more quite narrow and very high peaks ("lines" or "narrow bands")

then we cannot expect either to afford, or to be able to estimate, the great number of accurately chosen constants which would be required to obtain a rational function whose reciprocal has a shape very close ʼ ...
given narrow peak. We must adopt a slight ...
plan to make at least ...

... the peak. The
... ... of contribution from the peak and
... suitable for further prewhitening (if required) and analysis. (This operation can often, of course, be combined with further prewhitening so far as actual calculation goes. It will of course be necessary to compensate for the effects of this transformation at frequencies away from the peak, when preparing the final spectrum estimates for interpretation.)

There remains the estimation of the power in the peak, and possibly some inquiry into its width. A number of approaches are possible:

(1) We may analyze the original data as well as the data with the peak rejected, obtaining an estimate at the peak and possibly confirmatory estimates far from the peak.

(2) We may subtract a suitable multiple of the modified data from the original data so as to retain the peak and partially reduce other frequencies; and then analyze the difference.

(3) We may apply a band-pass filter to isolate frequencies at and near the peak, and then analyze the result.

Any of these techniques may be applicable in suitable circumstances.

Other related procedures are sometimes more natural than the use of moving linear combinations. Rejection of zero frequency, for example, is more naturally, and computationally more easily, accomplished by subtraction of the mean of all the data from each X_q than by the subtraction of a *moving* linear combination from each.

Rejection filtration has been applied in oceanography by Groves,[14] Seiwell,[15] Seiwell and Wadsworth,[16] to the elimination of various well-defined tides from records. It almost always has to be used to eliminate possible peaks at zero frequency (see Section 19 below).

In electronic measurements we may also anticipate its possible use in measurements: (i) close to a substantial harmonic of 60 cycles per second (such as 120 cps or 1380 cps), or (ii) near some strong "carrier".

17. SMOOTHING BY GROUPS

The cost of digital power spectrum analysis, once initial investments in programming, etc. have been made, and assuming records to have already been made and "read", is likely to be associated with the number of multiplications involved in computing the mean lagged products (in original or modified form). If there are n observations, and m lags are used, then there will be roughly nm multiplications.

Ways of reducing this number substantially are naturally of interest. Most of these must depend for their efficacy on our interest in something less than the whole spectrum. We have already discussed (in passing) a situation which would naturally arise only when we are interested only in the lower part of the aliased spectrum. This is the use of lags which are multiples of $\Delta\tau = h\Delta t$ with $h > 1$. The use of lags up to $m\Delta\tau = hm\Delta t$ allows us to explore the spectrum down to frequencies almost of the order $1/hm\Delta t$, which, had we used all multiples of Δt up to hm, would have required $hm + 1$ values of C_r (or of its modifications) instead of $m + 1$. The price of doing this is the aliasing of the spectrum with folding frequency $1/(2\Delta\tau) = (1/h)(1/(2\Delta t))$, which is h times as much aliasing as if all multiples of Δt up to hm had been used, yielding a folding frequency of $1/(2\Delta t)$.

If such intensive aliasing is bearable, this procedure with $\Delta\tau > \Delta t$ is simple, even though it is not necessarily economical. Indeed, if so much aliasing were permissible, we need only have "read" every hth data value. In many situations, however, especially where Δt has been taken as large as aliasing will permit, such further aliasing is unbearable. If we are to look at the low frequency part of the aliased spectrum $P_A(f)$ with computational economy, another course will have to be found.

Our use of linear schemes in prewhitening shows us a possible course. Let us begin by applying a linear scheme to the given values X_q, which attenuates all high frequencies. Then we can face further aliasing, and proceed apace.

If simplicity is controlling, then we take

$$\tilde{X}_q = X_q + X_{q-1} + \cdots + X_{q-k+1} \qquad (k \text{ terms})$$

for which the relation between the spectra (the power transfer function of the smoothing) is

$$\frac{\tilde{P}(f)}{P(f)} = |\, 1 + e^{-i\omega\Delta t} + \cdots + e^{-i(k-1)\omega\Delta t}\,|^2$$

$$= \left[\frac{\sin\dfrac{k\omega\Delta t}{2}}{\sin\dfrac{\omega\Delta t}{2}}\right]^2 .$$

This will give us zeroes at frequencies which are multiples of $1/k\Delta t$, and we can avoid folding the first two side lobes of this function onto the main lobe and still take a folding frequency as ~ ⸱⸱ choice will fold the second ⸳⸳⸳ lobe ⸳⸳ ⸱ ⸱

⸳⸳ ⸳⸳ be multi-

$$\left[\frac{\sin \dfrac{\omega \Delta t}{2}}{\sin \dfrac{k\omega \Delta t}{2}}\right]^2,$$

and only aliases which are usually negligible will have been superposed on the principal aliases. About one kth of the original principal spectrum will be available for analysis.

The stability obtained by this process can be easily compared with that obtained by using all X_q and taking $\Delta\tau = k\Delta t/4$. In each case, the width of the elementary frequency bands is approximately $1/2T'_n$ where T'_n has slightly different, but not substantially different values. The process just described yields nearly the same stability as $\Delta\tau = k\Delta t/4$, and usually involves much less computation, besides avoiding serious aliasing. It will almost always be preferred to using $\Delta\tau = h\Delta t$ with $h > 1$.

Other schemes of smoothing by groups are discussed in Section B.17.

18. PILOT ESTIMATION

The prewhitening procedure demands a rough knowledge of the spectrum for its effective use. Sometimes this rough knowledge can be obtained from theoretical considerations, or from past experience, but in many cases it must be obtained from a preliminary (pilot) analysis of the data. Such pilot analyses should be as simple and cheap as possible. We now discuss a pilot analysis giving very rough results quite easily.

Table III exemplifies a form of calculation which is easily carried out either entirely by hand, or with a desk calculator. The symbols "δ" and "σ" refer to differences and sums of consecutive numbers in non-overlapping pairs. Taking the numbers in non-overlapping pairs is not neces-

---

* Although this word should refer strictly to the deletion of only every 10th item, we shall apply it to the retention of only every jth item, for whatever j may be relevant.

Table III — Computation of Pilot Estimates

q	X_q	δX_q	$(\delta X_q)^2$	σX_q	$\delta\sigma X_q$	$(\delta\sigma X_q)^2$	$\sigma^2 X_q$	$\delta\sigma^2 X_q$	$(\delta\sigma^2 X_q)^2$
1	3								
2	4	1	1	7					
3	−1								
4	−2	−1	1	−3	−10	100	4		
5	2								
6	7	5	25	9					
7	5								
8	−1	−6	36	4	−5	25	13	9	81
9	−3								
10	2	5	25	−1					
11	5								
12	4	−1	1	9	10	100	8		
13	7								
14	3	−4	16	10					
15	4								
16	−1	−5	25	3	−7	49	13	5	25
17	−4								
18	2	6	36	−2					
19	4								
20	0	−4	16	4	6	36	2		
21	1								
22	−1	−2	4	0					
23	1								
24	2	1	1	3	3	9	3	1	1
25	4								
26	3	−1	1	7					
27	0								
28	−4	−4	16	−4	−11	121	3		
29	−1								
30	−2	−1	1	−3					
31	−2								
32*	−2	0	0	−4	−1	1	−7	−10	100
Totals			205			441			207

Continuation of Table III to the Right (Compressed)

q	$\sigma^3 X_q$	$\delta\sigma^3 X_q$	$(\delta\sigma^3 X_q)^2$	$\sigma^4 X_q$	$\delta\sigma^4 X_q$	$(\delta\sigma^4 X_q)^2$	$\sigma^5 X_q$	$(\sigma^5 X_q)^2$
8	17							
16	21	4	16	38				
24	5							
32	−4	−9	81	1	−37	1369	39	1521
			97			1369		1521

(* Note: 32 = 2^5.)

sary, but saves much calculation at little cost in accuracy. (In this table sums and differences are entered in the lower of the two lines to which they correspond.)

The final sums of squares are roughly proportional to the power in successive octaves coming down from the folding frequency. They differ by only a constant factor, equal to the number 2^t of values X_q used,

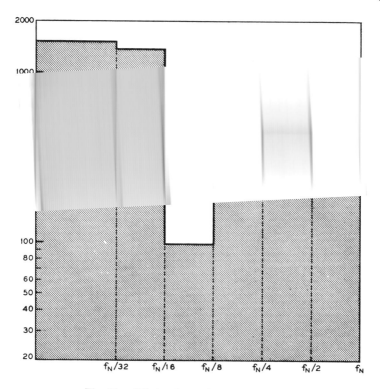

Fig. 11 — Pilot-estimated power spectrum.

from the mean squares of a nested analysis of variance. For many purposes they can be used as they come.

For the example of Table III we obtain sums of squares of 205, 441, 207, 97, 1369, and 1521. These are plotted in Fig. 11 for the successive octaves f_N to $f_N/2$, $f_N/2$ to $f_N/4$, $f_N/4$ to $f_N/8$, $f_N/8$ to $f_N/16$, $f_N/16$ to $f_N/32$, and the remaining range $f_N/32$ to 0. We see that the spectrum is roughly flat.

When medium or large stored-program digital computers are available, and the data is already available in machine-processable form (so-called diamond copy), it will often pay to use less elementary pilot calculations. Possible alternatives are discussed in Section B.18.

19. VERY LOW FREQUENCIES

The change from continuous "signals" processed in analog equipment to equally spaced "data" processed digitally has another important practical effect. Analog equipment, unless special care is taken, does not respond all the way down to zero frequency, and this automatically filters out the very lowest frequencies. This fact allowed us, in dealing with continuous records, to treat the "signals" being processed as if they had zero means. In dealing digitally with equally spaced data, all frequencies down to zero are transmitted, *unless* we take special precautions. Consequently, we must give serious attention to the very lowest frequencies.

(We must now distinguish between power (in the sense of a line) at zero frequency and power density at zero frequency. The power spectrum of a stationary random process with zero means may have finite power density at zero frequency without having finite power there. However, finite power at zero frequency may be introduced into the data in measurement. It would then be desirable to filter out the power at (exactly) zero frequency without affecting the power density at and near zero frequency due to the stationary random process, but this cannot be done perfectly.)

The need for such attention becomes clear when we consider the effect of "small" displacements of the average. Suppose that most of the observations (say about 999 in 1000) lie between -100 to $+100$, with a few falling outside one limit or the other. This would be the case when the standard deviation is about 30, the variance about 900. If the average of the observations were 5 or even 10, we might or might not detect at a glance its failure to be zero.

The total power is the square of the average (dc power) plus the variance. Numerically, perhaps $25 + 900 = 925$ or $100 + 900 = 1000$. All the dc power belongs to the very lowest frequency band, whose width is

$$\Delta f = \frac{1}{2T'_n}.$$

If we have data at one second intervals for a period of 15 minutes, a total of 900 points, we will have a folding frequency of one-half cycle per second, and 900 elementary frequency bands before we reach the folding frequency. Thus up to one tenth of all the power may be concentrated in one 900th of the spectrum, so that the lowest frequency band has a power density up to 90 times that of the average of the 899 others. It is not surprising that precautions need to be taken to deal with such possibilities. (After all, our standard spectral windows have side lobes more than 1 per cent the height of the main lobe.)

Slow trends, which may reasonably be regarded as zero-frequency sine waves, just as constant displacements are regarded as zero frequency cosine waves, are not nearly so likely to involve quite so substantial excesses of power density, but instances of this may and do arise.

Any way of dealing with these effects must essentially remove the lowest elementary frequency band, or both this band and the next to lowest one. In the process it will also have to eliminate some parts of the next higher elementary bands as well, since we cannot d̲e̲s̲i̲̲ ̲̲̲̲̲

procedure entirely f̲r̲o̲o̲ ̲o̲f̲ ̲.̲.̲.̲ ̲.̲.̲

, ̲o̲.̲.̲o̲v̲o̲i̲, be summarized as apply-

̲.̲ ̲.̲.̲.̲ ̲.̲.̲i̲i̲t̲o̲ cosine series transform to

$$C_r - E_{kr}$$

where k identifies a specific method of modification, rather than to the C_r alone.

In place of

$$\text{ave } \{P_{iA}(f)\} = Q_{iA}(f) * P_A(f),$$

we shall now have

$$\text{ave } \{P_{iAk}(f)\} = Q_{iAk}(f) * P_A(f) = [Q_{iA}(f) - R_{ik}(f)] * P_A(f)$$

where $R_{ik}(f)$ is related to the E_{kr} in the same way that $Q_i(f)$ is related to the C_r.

Details for certain special choices for E_{kr} are given in Section B.19. It is there concluded that, among others, satisfactory choices for practical calculation appear, for the present, to be, for removing possible constants,

$$E_{0r} = (\bar{X})^2 \qquad \text{(independent of } r)$$

and, for removing the effects of both possible constants and possible linear trends,

$$E_{1r} = (\bar{X})^2 + \frac{3}{16}\left(1 - \frac{1}{n^2} - \frac{2r}{n} - \frac{2r^2}{n^2}\right)(\overline{X^+} - \overline{X^-})^2$$

where $\overline{X^+}$ and $\overline{X^-}$ are the means of the right- and left-hand thirds of the X values.

WARNING: It will *almost never be wise* to fail to use some E_{kr} in a digital computation.

ANALYSIS IN PRACTICE

The two sections which follow discuss the questioning and planning required whenever a digital analysis of equally spaced data is to be made, and exhibit a sample sequence of calculation formulas which might result from such planning. They are intended to summarize the previous material in its application to analysis. (Application to planning for measurement is treated next after this.)

20. PRACTICAL ANALYSIS OF AN EQUALLY SPACED RECORD

We may logically and usefully separate the analysis of an equally spaced record into four stages — each stage characterized by a question:

(a) Can the available data provide a meaningful estimated spectrum?

(b) Can the desires of the engineer for resolution and precision be harmonized with what the data can furnish?

(c) What modifications of the data are desirable or required before routine processing?

(d) How should modification and routine processing be carried out?
Failure to adequately consider any one question properly, or failure to apply any one answer, can make the entire analysis worthless.

The data presented will have come about by measuring some physical phenomenon at regular intervals. Thus,

1. the spectrum of the phenomenon

2. the frequency response of the instruments used to make the measurements

3. the probable magnitudes of measuring, and recording or reading errors, and

4. the time separation between adjacent values
are all relevant.

The first stage of consideration is to inquire generally about these quantities, and to determine whether either aliasing (see Section 12) or background noise is so heavy as to make the values almost wholly useless. Thus, if the spectrum is believed to extend up to 10 megacycles with substantial intensity, if the measuring equipment is flat to 1.2 kilocycles and is 60 db down at 5 kilocycles, and if the values are measured every $\frac{1}{200}$ of a second, we may as well stop here and go no further, since the whole available spectrum (up to 100 cycles) will be aliased more than a dozen times over. (The 1.2 kilocycle measurement bandwidth, which will be aliased 12 layers deep, will control rather than the 10 megacycle phenomenon bandwidth.)

If, on the other hand, the equipment was flat to 10 cycles, down about

6 db at 20 cycles, 15 db at 30 cycles, and 60 db at 50 cycles, we would not expect any irremovable aliasing difficulties, and would expect to be able to estimate the spectrum up to some moderate f~~
say. 20 cycles. 30

...... and converted into the approximate number of elementary frequency bands (number of degrees of freedom — see Section 9 which is based on Sections 6 to 8) possibly available for each of the proposed estimates. This number can then be compared with the number of degrees of freedom required (also see Section 9) to give the desired fractional accuracy. If these are consistent, or if the desired accuracy, or the desired resolution, or both can be modified to make them consistent, then there is a good chance that the data can be persuaded to yield the desired results, and further inquiry is indicated. If not, we should stop here.

Explicit relations among duration, resolution, and fractional accuracy, the latter expressed in terms of 90 per cent interval (cp. Tables I and II), are given in Section B.23. These lead to an approximate 90 per cent spread, expressed in db (decibels), of

$$\frac{14}{\sqrt{\text{(total duration in secs)(resolution in cps)}} - \frac{1}{2} - \frac{1}{3} \text{ (number of pieces)}}$$

a result which may often be conveniently used in such an inquiry.

At the beginning of the third stage, information should be sought as to

1. over what range of frequencies the spectrum is desired, and

2. whether any lines or high and narrow peaks are to be expected, and at what frequencies.

Guided by this information, it should be possible to decide whether either

a. smoothing by groups (as in Section 17) to reduce computation without loss of low-frequency information, or

b. rejection filtration (as in Section 16) to suppress well-established lines or high and narrow peaks,

or both, are desirable. If desirable, they are then carried out before or during the next step.

Unless advance information about the spectrum is exceedingly good, a pilot analysis (see Section 18) to establish the rough form of the spectrum will now be very much worthwhile. The result (or the very good advance information, if available) will now make it possible to choose a reasonable prewhitening procedure (or, possibly, to choose not to prewhiten). Once suitable prewhitening (see Sections 11 and 15) has been chosen, and either carried out or planned for, the third stage is complete.

Finally, the information on resolution and accuracy combine to specify the width of spectral window desired, and hence (see Section 13) the number of lags for which mean lagged products should be calculated. When these are in hand, they are modified and transformed (or, perhaps more simply, transformed and convolved — see Section 13), adjusted to screen out very low frequencies, and the resulting power density estimates are corrected for the prewhitening, and for grouping and/or rejection filtration (if any) used. The final estimates are best plotted on a logarithmic power scale, since their accuracy will be roughly constant on this scale. Crude confidence limits can then be calculated from the number of degrees of freedom (see Section 9) which would be present in the individual estimates if: (i) the process were Gaussian, and (ii) the prewhitened spectrum were flat. (The factor of safety of Section 8 will ordinarily be adequate.)

21. SAMPLE COMPUTING FORMULAS

We cannot prescribe one set of computing formulas for general use, since there are rational reasons for different choices. All we can do is illustrate a procedure which may work fairly well in many cases. (And our example is not likely to be the only one with such properties. If the reader understands, by comparison with adjacent sections, just why we do what we do, he can compare other procedures with this example in a meaningful way. He will have to understand much of what is said in order to do this.)

If X_t, $t = 0, 1, \cdots, n$ are the given observations, which we will treat as if at unit spacing, it is likely that $P_A(f)$ decreases substantially as f goes from 0.0 to $0.5 = f_N$. (If it does not, then aliasing is likely to have been serious, and satisfactory analysis at this spacing may be impossible.) Prewhitening by

$$\tilde{X}_t = X_t - 0.6\, X_{t-1}$$

which multiplies $P_A(f)$ by

$$1.36 - 1.20 \cos 2\pi f,$$

a factor increasing from 0.16 to 2.56, *may* be a wise prewhitening. (The index t will now start at 1, and not at zero.)

We calculate next

$$\left[V_0 + 2 \sum_{q=1} C'_q \cdot \cos \frac{qr\pi}{m} + C'_m \cdot \cos r\pi \right]$$

and the results of hanning (see Sections 5 and 13)

$$U_0 = \tfrac{1}{2}(V_0 + V_1)$$
$$U_r = \tfrac{1}{4}V_{r-1} + \tfrac{1}{2}V_r + \tfrac{1}{4}V_{r+1}, \qquad 1 \leqq r \leqq m - 1,$$
$$U_m = \tfrac{1}{2}V_{m-1} + \tfrac{1}{2}V_m .$$

These can then be corrected for both prewhitening and the correction for the mean by forming (see Section B.21)

$$\frac{n}{n - m} \frac{1}{1.36 - 1.20 \cos \dfrac{2\pi}{6m}} U_0 ,$$

$$\frac{1}{1.36 - 1.20 \cos \dfrac{2r\pi}{2m}} U_r , \qquad\qquad 1 \leqq r \leqq m - 1,$$

$$\frac{1}{1.36 - 1.20 \cos \left(1 - \dfrac{1}{6m}\right) 2\pi} U_m ,$$

as smoothed estimates of the power density. Estimates with subscript 0 will apply in the range just above zero frequency, those with subscript r near a frequency of $r/(2m)$ cycle per observation, and those with subscript m in the range just below a frequency of 0.5 cycle per observation.

In interpreting these estimates four cautions are important:

(a) aliasing of frequencies (see Section 12) may have taken place,

(b) the estimates are smoothed with a crudely isosceles triangular weighting function (see Sections 5 and 13) of full width $4/(2m)$,

(c) no estimate will be more stable than chi-square on $(2n)/m$ degrees of freedom and, wherever the spectrum is not smooth, the stability of the estimates will be appreciably less (see Section 9),

(d) adjacent estimates will not have independent sampling errors, though those not adjacent are at least very close to being uncorrelated.

The units involved are such that the smoothed one-sided, aliased power density on $0.0 \leqq f \leqq 0.5$ is approximated by twice the estimates. The pieces into which the variance would be divided, each coming from a frequency band of width $1/(2m)$ cycles per observation, are estimated by $1/(2m)$ times the corrected estimates.

Planning for Measurement

Up to this point, with the exception of part of Section 11, our discussion has been concerned (i) with what happens when certain operations are performed, and hence (ii) with how we should make the best of what we already have.

The third aspect — planning the measurements or observations to meet requirements — has not been adequately treated. (Both statisticians and engineers concerned with measurement will agree that this is the most vital aspect of all, but will, unfortunately, also have to admit that, all too often, "salvage" work will be required because this third aspect was omitted, and the observations made unwisely.)

In discussing "What data shall we take?", "How shall we measure it?", the same considerations will recur as in discussing "How shall we analyze it?", but (i) they will be looked at from quite different aspects and (ii) they will be even more important. Now, by planning in advance of data-gathering, we may be able either to replace useless or difficult-to-analyze measurements by usable ones, or to avoid making measurements which could never provide the desired information.

The first basic decision has to do with the type of recording and analysis to be used. Three types are in use today:

(1) *Spaced:* Analog use of intermittent recorders (photography of situations or of dials, etc.) or digital recording at equally spaced intervals (electronic reading of dials, photography of counters, etc.).

(2) *Mixed:* Continuous recording (on film, calibrated paper rolls, etc.) with the intention of analyzing equally spaced values to be "read" from these continuous records.

(3) *Continuous:* Continuous recording (FM recording on magnetic tape, etc.) with the intention of making an analog analysis.

The choice among these types will depend on their particular advantages and disadvantages, and on the availability of equipment, both for recording and analysis. In almost every case, however, the ... ... problems will be ...

... ... is so ... ... read, the high-frequency response may indeed be reduced by filtering. Such filtering is too likely to be regarded as unfortunate rather than helpful. Effort tends always to be applied for "faithful" recording. This is appropriate for recording specific *individual* time histories for *visual* study, but is often *most. inappropriate* for recording *sample* time histories for *statistical* study with the aid of sensitive *filters* (analog or digital). (When the recording is continuous, be it on film, oscillograph paper, or magnetic tape, the "writing" means has a limited frequency response, and this will usually help to keep the record from blurring.)

When the analysis is to be made on equally spaced data, whether the recording be continuous or equi-spaced, there is a real problem of aliasing. And there is need for a basic choice of a frequency cutoff, usually in terms of two frequencies such that (i) the experiment is only concerned with frequencies up to the lower one, and (ii) frequencies beyond the upper one will not be recorded. The need for such a choice in a continuous system may not appear to be so acute, since only problems of noise or non-linear distortion are involved (see Section 11). Yet in practice, it will almost always be made — indirectly — by the choice of a writing speed (which implies a frequency cutoff for a continuous recorder). Economic pressures to reduce both the volume of record, and the extent of measurement and computation, act to lower the frequency cutoff, while desires to follow the spectrum to higher frequencies act to raise it. The proper choice comes from balancing these pressures.

Sometimes in mixed systems, when continuously recorded data is to be subjected to equi-spaced analysis, an attempt is made to compromise matters by recording with a high cutoff, and then asking that the measurements of this record be "eye averages" over periods long enough for the record to show considerable variation. Such compromises do not seem to work nearly as well in practice as their proponents suppose. Replacing the "averages" by the results of "reading to the line" at equi-spaced points often seems to give better results, even though a smaller, but unknown amount of aliasing is thus replaced by a larger, known amount. Putting the filtering into the observing and writing equipment, rather than into the (human) measurer and transcriber, will usually do even better — better by a large margin.

If one can be confident of the upper limit, beyond which the power spectrum will not be needed, it is usually best to record with a related frequency cutoff, thus reducing noise complications, aliasing difficulties, and the necessary bulk of the record.

Conversely, however, points must be recorded or measured frequently enough (or a high-enough writing speed used) so that aliasing (or loss of high-frequency response) is not serious. (For a given maximum usable frequency, the sharper the cutoff, the less stringent this requirement.)

To summarize, the problems surrounding aliasing should lead to the choice of a frequency cutoff which is usefully described by two frequencies (which may reasonably be in the ratio of 1 to 2):

(a) a lower frequency, which is the highest at which important power spectrum estimates will be made, and

(b) a higher frequency, at and above which no serious amount of recording is done.

Both of these need to be chosen before settling finally on observing and recording equipment. If equi-spaced data is produced, the folding frequency may be as low as half-way between these two frequencies.

A prime essential to keep in mind is that all measurement, transmission, *and analysis* systems are essentially band-limited. It is always inadvisable to try to cover too many octaves of log frequency while using exactly the same techniques.

23. DURATION OF DATA REQUIRED

Instead of trying to compromise resolution and stability within the limitations of available data, we may now consider the costs and advantages of getting still more data, or, perhaps, somewhat less data. We face a three-way compromise among effort, resolution, and stability (precision) of estimate.

Effort has to be measured in various ways, but the duration of initial record will almost certainly have to be considered as one measure. It is shown in Section B.23, where both precise definitions of the quantities, and a corresponding formula for the necessary numbers of pieces of a given length will also be found, that

$$\text{(total duration in seconds)} = \frac{\dfrac{1}{2} + \dfrac{200}{(90\% \text{ range in db})^2} + \dfrac{\text{(pieces)}}{3}}{\text{(resolution in cps)}}.$$

... processing of spaced records, so long as we apply the best methods which we know to a shape of spectrum which is not exceptionally difficult to handle.

24. AMOUNT OF DIGITAL DATA-HANDLING REQUIRED

If spaced data are to be digitally processed, both the number of data points to be used and the number of multiplications involved are of interest.

If we can easily build in the desirable frequency cutoff, and have to resolve a number of contiguous uniformly narrow bands from zero frequency to some maximum frequency, then we will require about

$$\left[\frac{3}{2} + \frac{600}{(90\% \text{ range in db})^2} + \text{(pieces)}\right] \text{(number of bands resolved)}$$

data points and, roughly about

$$\left(\frac{9}{2} + \frac{1800}{(90\% \text{ range in db})^2} + 3 \text{ (pieces)}\right) \text{(number of bands resolved)}^2$$

multiplications.

These last two results often give only preliminary indications. Aliasing difficulties will increase these numbers. The possibility of smoothing by groups will decrease them. Details and possible modifications of the proposed system of data gathering and analysis need to be studied carefully before final estimates of the number of data points and the rough number of multiplications are finally settled upon.

25. QUALITY OF MEASUREMENT AND HANDLING

In every case, careful consideration should be given to the quality of measurement and data handling required (in terms of the dangers of, e.g.: time-varying frequency response, introduced noise, intermodulation distortion, etc.). An extensive catalog would be out of place here, since the problems are basically those of instrumentation engineering. But a few reminders may indicate the diversity of problems which might arise. A camera may be "clamped" to some object to record the relative orientation of that object and something visible to the camera. The mounting of the camera is never perfectly rigid, and vibrations will occur ordinarily at frequencies far above the data-taking rate. Whatever the frequency, these vibrations will introduce "noise" into the record. At least an order-of-magnitude calculation of the effects of likely vibration is needed.

Storage of a signal on magnetic tape will be a part of many measurement-analysis systems. Because only rough spectra are wanted, AM (amplitude modulation) recording may be planned. If the fact that AM recording and playback is subject to considerable fluctuation in over-all gain (db's, not tenths db) is neglected, measurement planning may be quite misleading.

In a complex analysis, where several spectra and cross-spectra (whose analysis we have not specifically discussed) are involved, it might be planned to plot the estimates of each spectrum and cross-spectrum against frequency, draw smooth curves, and compute derived quantities from values read from these curves. Such a process has led to great difficulties in certain actual situations, because of the "noise" introduced by such visual smoothing which appears to have distinctive but unknown properties. Such a graphical step may appear to be good engineering, but it cannot be high quality data handling. Its use may nullify the careful selection of other data processes, some of which are delicately balanced.

Graphical analysis should ordinarily be reserved for:

(a) display of whatever spectrum or function of spectra is really a final output,

(b) description of the actual effects of computational procedures, and

(c) trouble-shooting.

26. EXAMPLE A

Suppose first that the spectrum of some aspect of the angular tracking performance of a new radar is to be obtained; that angular tracking can only be studied by photographing the target with a camera clamped to the antenna; that frequencies near 0.27 cps are of special interest; that the spectrum of tracking performance at higher frequencies is relatively flat up to 10 cps and then falls rapidly enough to be negligible beyond 40

cps; that estimates at all frequencies up to 25 cps are desired; and that
stability to $\pm$ 1 db is desired. What are the requirements?

The total amount of tracking required is fixed by the ~~~ ' ''
quirement near 0.27 cps which
0.00

..ovomes or either 16 or 40 minutes continuous tracking are almost certain to be out of the question. The length of piece available
would depend on the aspect of tracking performance studied, but a fair
figure for this illustration might be 200 seconds. Going to Section B.23
for the necessary formula, we find

$$\text{(number of pieces)} = \frac{\frac{1}{2} + 50}{(200)(0.05) - \frac{1}{3}} = \frac{50.5}{9.67} = 5.2$$

or

$$\text{(number of pieces)} = \frac{\frac{1}{2} + 50}{(200)(0.02) - \frac{1}{3}} = \frac{50.5}{3.67} = 13.7.$$

From a purely experimental point of view, these amounts of data are
moderately hard to substantially hard to obtain, but we may suppose
them available as far as radar and target availability are concerned.

We come next to data taking and availability problems. We must
study the spectrum up to 25 cps. Since the spectrum is negligible only
above 40 cps, our folding frequency must be at least 32.5 cps, which
would fold 40 cps exactly back to 25 cps. Hence we need at least 65
frames a second. Consideration of available frame rates bring us to 64
frames a second as probably reasonable. This is 12,800 frames in each 200
second piece, a total film reading load of between 50 and 150 thousand
frames. This will require some hundreds of man-days of film reading,
but may perhaps be faced.

To calculate *directly* the rough number of multiplications involved, we may begin by assuming that we are going to require the 0.05 or 0.02 cps resolution all the way from 0 to 25 cps. Were this the case, then we would require to resolve from

$$\frac{25}{0.05} = 500$$

to

$$\frac{25}{0.02} = 1{,}250$$

frequency bands. The corresponding numbers of multiplications range from

$$[4.5 + 450 + 3 \text{ (pieces)}] (500)^2 \approx 120 \text{ million}$$

to

$$[4.5 + 450 + 3 \text{ (pieces)}] (1{,}250)^2 \approx 750 \text{ million.}$$

The running time of an IBM 650 calculator on such a problem is about 10 hours per million multiplications, so that between

$$1{,}200 \text{ hours} = 30 \text{ shift-weeks}$$

and

$$7{,}500 \text{ hours} = 188 \text{ shift-weeks}$$

would be required. Clearly these machine times are out of line, and attention should be given to ways of reducing this aspect of effort.

An application of smoothing by groups seems most likely to be effective, especially since the high resolution is only wanted near the low frequency of 0.27 cps. Let us suppose that, in view of the supposed rather flat spectrum out to 10 cps, the engineers concerned will be content with two spectrum analyses, one with 0.5 cps resolution extending all the way to 25 cps, and the other with 0.02 cps resolution extending only to 1 cps. What effect will this have on the computational load?

Notice first that it will have no effect on the radar-and-target operating and film-reading loads. These were fixed by the resolution-precision requirements, and by the combination of this with the upper limit of the actual spectrum affecting the camera. Replanning details of the analysis will save nothing on either of these.

The broad-frequency low-resolution analysis will resolve about

$$\frac{25}{0.5} = 50 \text{ bands}$$

and require roughly

$$[4.5 + 450 + 3(14)] \cdot (50)^2 = 1.24 \text{ million multiplications}$$

(since we shall need 14 pieces to obt~~~~~

~~~~~~~ sis will resolve

$$\frac{1.0}{0.02} = 50 \text{ bands}$$

and will require about another 12.4 hours of machine time.

Thus we have reduced machine time to about 25–30 hours, in pleasant contrast with the remaining requirements of some hundreds of hours of film reading and 14 test runs of 200 seconds each. The balance is approximately restored.

Our apparently blind use of the multiplications-required formula has concealed one important point. Our calculation of the time required for the high-frequency low-resolution analysis tacitly assumed that we have processed no more of the data than is required to meet the actual resolution-precision requirement.

The loosening of resolution from 0.02 cps to 0.5 cps in this part of the analysis has reduced by a factor 25 the amount of data which must be processed to meet the ±1 db (90 per cent) requirement. Hence the two hours machine time is predicated on processing only $\frac{1}{25}$th of the available data. If only about $\frac{1}{25}$ of the data is to be processed for the high frequency analysis, then it will be desirable to take the most typical 8 or 10 seconds from each piece. The losses due to end effects will be somewhat greater, it is true, but the advantages of increased coverage of the effects of unplanned variation, consequent on using parts of all 14 runs, far outweigh such considerations.

It would be possible to use only one run for the high-frequency analysis, a possibility which emphasizes the fact that $\frac{13}{14}$ths of the film reading is done to obtain the raw material for averaging, for filtering out high frequencies. If the hundreds of man-days of film reading look out of line, *and if* the line from the radar to the target is known not to change rapidly (with respect to an inertial frame of reference), then we are driven to consider whether the "clamping" of the camera to the antenna could be modified in such a way as to provide a frequency cutoff between antenna position and camera position. What would be desired would be a reliable mechanical filter with a cutoff at 1 or 2 cps, and substantial, reproducible transmission up to, say, 0.5 cps. If such a mount could be taken down from the shelf, then it would suffice to make (a) one 200-second run with a stiff mount and 64 frames per second, and, say, (b) thirteen 200-second runs with a mount of such designed softness, and, say 4 frames per second. The total number of frames for reading would now be 12,800 for run (a) and 800 for each run (b), a total of about 23,000 frames. This might require about a man-month to read, a saving of several man-months. Unfortunately, such a sharply-tuned low-pass mount would not be likely to be on the shelf.

27. EXAMPLE B

As a second example, suppose a new solid-state device develops a noise voltage with a power spectrum roughly proportional to $1/f^2$ when under test under most extreme circumstances — circumstances so extreme that its average life is 30 to 50 milliseconds, and suppose that the detailed behaviour of this spectrum is believed likely to provide a clue to the proper theoretical treatment of some of the properties of this device. Suppose further that, while it was believed that the shape of the spectrum of the noise from different examples of this device was the same, the voltage levels of different devices were quite different. It might be reasonable to ask for spectral measurements to ± 0.25 db resolving 1 cps and covering from 1 cps to 500 cps. Direct measurements are likely to be most difficult, for the power between 499 and 500 cps is about $\frac{1}{120,000}$th the power between 1 and 2 cps, a difference of 51 db in level. Our recording and processing equipment is not likely to have the dynamic range required for direct analysis.

Clearly we should prewhiten our noise as early in the measurement and analysis system as we reasonably can. Fortunately, prewhitening here is

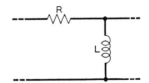

Fig. 12 — RL voltage divider.

operationally simple. A RL voltage divider, as indicated in Fig. 12 will introduce an attenuation of voltage, if the load impedance is high, amounting to

$$| R + i\omega L |^2$$

$$\frac{\overline{\omega^2}}{1 + \dfrac{R^2}{\omega^2 L^2}} = \frac{4\pi^2 A L^2}{R^2 + \omega^2 L^2}$$

which will be initially constant, and then decrease 6 db per octave, with a corner at $\omega_c = R/L, f_c = R/2\pi L$. As a first step in measuring a spectrum out to, say, $f = 2R/2\pi L$, at which frequency the prewhitened spectrum would be down about 7 db, such a change would be useful. The range of frequencies which could be usefully studied would not be appreciably reduced by such a change, even though the low frequency power level would be greatly reduced by the prewhitening network, since the low-frequency power level would not be seriously reduced below the former power level at the corner frequency. If one could have been studied, the other can be studied.

28. EXAMPLE C

The irregularities in the earth's rotation have been studied by Brouwer,[17] who reduced the available observations (times of occultation and meridian passage) by averaging over individual years. He states "occultations so reduced in recent years have been demonstrated to yield annual means essentially free from systematic errors if the observations are well distributed over the year. ... The δ's may themselves be the accumulations of numerous smaller random changes with average intervals much smaller than a year. The astronomical evidence throws no further light on this, though perhaps something may be gained by an analysis of residuals in the moon's mean longitude taken by lunations."[18] These comments suggest that astronomical data can supply values once a year, possibly no more frequently, and may be able to supply values about 13 times a year (once per lunation), certainly no more frequently. Let us accept the first possibility as a basis for an example. (This is the best example we know of a situation where equally spaced data cannot, in principle, be had at a finer spacing.)

The information most nearly directly supplied by the astronomical observations is Δt, the difference between ephemeris time and mean solar time. Brouwer discusses two statistical models for its structure, both of which are most easily described in terms of the behavior of the second differences of the observations. In the first, the true second differences are constant over periods of varying length. In the second model, the true second differences are independently and randomly distributed. In either case, observational errors, independent from observation-period to observation-period also contribute to the observed Δt's.

If we were to plan an observational program to decide between these hypotheses by spectral analysis we need first to specify the alternative spectra. The first model seems never to have been made as precise statistically as the second. Brouwer's fitted curves correspond to constancy over periods of from 4 to 15 years. We should like to get a general idea of the possible spectra corresponding to this model without making the model too specific. Consider first a situation in which, except for the effects of second differences of experimental errors, the observations are constant in blocks of five, and where the values assigned to different blocks are independent. The successive average lagged products (starting with lag zero) are proportional to 5, 4, 3, 2, 1, 0, 0, 0, . . . and it follows that the power density is proportional to

$$1 + \frac{8}{5} \cos \pi f/f_N + \frac{6}{5} \cos 2\pi f/f_N + \frac{4}{5} \cos 3\pi f/f_N + \frac{2}{5} \cos 4\pi f/f_N .$$

Calculation shows that this is high near zero frequency, falling rapidly until, beyond about $f/f_N = 0.3$, it consists of ripples with an average height of less than $\frac{1}{25}$th the low frequency peak. If, instead of "constant by fives", the specific model were "constant by eights" or "constant by tens", still with independence between blocks, this peaking would be more pronounced and confined to still lower frequencies. If the lengths of the blocks were to vary at random, according to some distribution, still with independence of value, the spectrum would be the corresponding average of such spectra for fixed block lengths. The spectrum to be expected *for second differences* of annual average observations then should consist of a sum of two components:

(1) a "true" component peaked at low frequencies, falling rapidly by, say, $f/f_N = 0.2$ or 0.25, and continuing to $f/f_N = 1.00$ with an average height perhaps 1 per cent or 2 per cent of the low-frequency value,

(2) an "observational error" component, corresponding to independent errors in the annual averages, and hence proportional to $(1 - \cos{(\pi f/f_N)})^2$.

In case the second model should apply, the first compo~~~ replaced by one with ~ ~ ~

.....range density should be ~ ~ somewhat greater than the low-range density, the increment representing effects of observational error.

Without more detailed estimates of the relative sizes of the components, it would be difficult to specify exactly how many observations would be required to separate Model I from Model II, but 10 to 20 degrees of freedom in each of the ranges discussed should be quite helpful. This suggests 100 values of annual second differences, corresponding to 102 years of careful astronomy, as likely to be helpful. Since Brouwer gives annual values for 131 years, some 129 annual second differences are

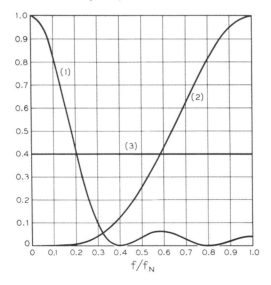

Fig. 13 — Components for two models of earth-rotation irregularities: (1) "true irregularity" component for first model, (2) "observational error" component for either model, (3) "true irregularity" component for second model.

available for trial, and it may be possible to answer the question without waiting for many more years to pass.

It might well suffice to estimate smoothed densities over octaves such as $0.0625 \leqq f/f_N \leqq 0.125$, $0.125 \leqq f/f_N \leqq 0.25$, $0.25 \leqq f/f_N \leqq 0.5$ and $0.5 \leqq f/f_N \leqq 1$. Thus we might consider using the add-and-subtract pilot estimation method for initial exploration. The actual analysis of Brouwer's data is considered further in Section B.28.

APPENDIX A

FUNDAMENTAL FOURIER TECHNIQUES

In this appendix we review briefly certain aspects of Fourier transformation. These aspects may be regarded as dealing mainly with diffraction by slits, rectangular or graded, and by analogs made up of discrete "lines". Convolution and the so-called Dirac functions are specially important as convenient tools. Some parts of the discussion will have no direct bearing on the analysis of procedures for power spectrum estimation, but are intended to familiarize the reader with analytical tools which are used frequently throughout the remainder of this paper, and which may be used to advantage in many other analyses of a similar nature.

A.1 *Fourier Transformation*

There are several formulations of Fourier transformation which differ according to custom, convenience, or taste. The formulation which we will adopt here is the one used by Campbell and Foster.[19] Given a function of time, $G(t)$, its Fourier transform is a function of frequency, and is given by the formula

$$S(f) = \int_{-\infty}^{\infty} G(t) \cdot e^{-i\omega t}\, dt \qquad (\omega = 2\pi f).$$

Conversely, given a function of frequency, $S(f)$, its Fourier transform is a function of time, and is given by the formula

$$G(t) = \int_{-\infty}^{\infty} S(f) \cdot e^{i\omega t}\, df \qquad (\omega = 2\pi f).$$

The term "frequency" is used here, not in the probability or statistical sense, but in the sense of sinusoidal or cisoidal functions of time ($\cos \omega t$, $\sin \omega t$, $e^{i\omega t}$).

Our preference for the Campbell-Foster formulation is based on the following points, arranged approximately in the order of increasing weight.

1. Frequencies are expressed in cycles per second more naturally and much more frequently than in radians per ~~~~ ¹ ⁻
use ω only ~~ ~~ ¹ ¹

$$G(t) = \frac{1}{2\pi i} \int_{-i\infty}^{i\infty} S\left(\frac{p}{2\pi i}\right) \cdot e^{pt} \, dp$$

is a natural and convenient step in the calculation of the integral by the method of residues.

4. The transformation formulae correspond to the conventional relations between the *impulse response* (response due to a unit impulse applied at $t = 0$) and the *transfer function* (ratio of steady-state response to excitation, for the complex excitation $e^{i\omega t}$) of a fixed linear transmission network. These network functional relations are commonly regarded as Laplace transformations rather than Fourier transformations. As a matter of fact, however, the circumstances in almost all practical applications are such that there is no essential difference between Laplace transformations and Fourier transformations. Impulse responses are zero for $t < 0$ and vanish exponentially as $t \to \infty$, and transfer functions are analytic on and to the right of the imaginary axis (including the point at infinity) in the complex p-plane. On the very rare occasions when a communications engineer might be interested in the behavior of a network under energetic initial conditions, he has ways of introducing the initial conditions without using Laplace transforms (Guillemin[20]).

It should be noted that, since $G(t)$ must be a real function, the real part of $S(f)$ must be an even function, and the imaginary part of $S(f)$ must be an odd function. The even part of $G(t)$ and the even (real) part of $S(f)$ are cosine-transforms of each other. The odd part of $G(t)$ and the odd (imaginary) part of $S(f)$ are negative sine-transforms of each other. It should be noted also that if $G(t)$ and $S(f)$ constitute a transform-pair, then $G(-t)$ and $S(-f)$ also constitute a transform-pair. Further, $S(-f)$ is equal to $S^*(f)$, the complex conjugate of $S(f)$.

A.2 *Some Transform-Pairs*

We will now turn our attention to some transform-pairs which we will require directly or indirectly in the analysis of procedures for power spectrum estimation. We will use special symbols for some of these transform-pairs. For later reference, these transform-pairs will be collected in Table IV.

The first transform-pair, which is easily worked out, involves a symmetrical rectangular time function (box car of length $2T_m$), viz.

$$D_0(t) = 1, \qquad |t| < T_m,$$
$$= \tfrac{1}{2}, \qquad |t| = T_m,$$
$$= 0, \qquad |t| > T_m.$$

TABLE IV

1.	$D_0(t) = 1,\ |t| < T_m$ $= \dfrac{1}{2},\ |t| = T_m$ $= 0,\ |t| > T_m$	$Q_0(f) = 2T_m \dfrac{\sin \omega T_m}{\omega T_m}$ $= 2T_m \ \mathrm{dif}\ 2fT_m$
2.	$D_1(t) = 1 - \dfrac{|t|}{T_m},\ |t| \le T_m$ $= 0, \qquad |t| \ge T_m$	$Q_1(f) = T_m \left(\dfrac{\sin \pi f T_m}{\pi f T_m}\right)^2$ $= T_m\ (\mathrm{dif}\ \dot{f} T_m)^2$
3.	$\delta(t - t_0)$	$e^{-i\omega t_0}$
4.	$\cos \omega_0 t$	$\tfrac{1}{2}\ [\delta(f + f_0) + \delta(f - f_0)]$
5.	$\nabla_m(t\,;\Delta t) = \dfrac{\Delta t}{2}\,\delta(t + m\Delta t)$ $+ \Delta t \cdot \sum\limits_{q=-m+1}^{q=m-1} \delta(t - q\Delta t)$ $+ \dfrac{\Delta t}{2}\,\delta(t - m\Delta t)$	$Q_0(f\,;\Delta t)$ $= \Delta t \cdot \cot \dfrac{\omega \Delta t}{2} \cdot \sin m\omega \Delta t$ $= 2(m \cdot \Delta t)\cos(\pi f \cdot \Delta t)\dfrac{\mathrm{dif}\ 2f(m\cdot \Delta t)}{\mathrm{dif}\ f\cdot \Delta t}$
6.	$\nabla(t\,;\Delta t) = \Delta t \cdot \sum\limits_{q=-\infty}^{q=\infty} \delta(t - q\Delta t)$	$A\left(f;\dfrac{1}{\Delta t}\right) = \sum\limits_{q=-\infty}^{q=\infty} \delta\left(f - \dfrac{q}{\Delta t}\right)$
7.	$A(t\,;\Delta t)$	$\nabla\left(f\,;\dfrac{1}{\Delta t}\right)$

The corresponding frequency function is

$$Q_0(f) = 2T_m \frac{\sin \omega T_m}{\omega T_m} = 2T_m \cdot \text{dif } 2fT_m.$$

(The ~~~~

~~~~-pair, which is almost as readily worked out as the first, involves a symmetrical triangular time function, viz.

$$D_1(t) = 1 - \frac{|t|}{T_m}, \qquad |t| \leq T_m,$$

$$= 0, \qquad\qquad |t| \geq T_m.$$

The corresponding frequency function is

$$Q_1(f) = T_m \left(\frac{\sin \pi f T_m}{\pi f T_m} \right)^2 = T_m (\text{dif } f T_m)^2.$$

Except for scale factors, this frequency function behaves as shown in Fig. 14.

The third transform-pair involves a so-called Dirac *function* as the time function. The Dirac function is not a function in the strict mathematical sense. It is called a "measure" by L. Schwartz.[21] For our purposes, it will only be necessary to identify $\delta(t - t_0) \cdot dt$ formally with $dh(t - t_0)$ where $h(t - t_0)$ is Heaviside's unit-step function, viz.,

$$h(t - t_0) = 0, \qquad t < t_0$$

$$= 1, \qquad t > t_0$$

and to interpret all integrals as Stieltjes integrals. Hence if the time function (to use the term loosely) is

$$G(t) = \delta(t - t_0)$$

then, the corresponding frequency function is

$$S(f) = e^{-i\omega t_0}.$$

It should be noted that while $\delta(t - t_0)$ is easily formally transformed into a frequency function, the latter is not so readily transformed into the original time function.

The fourth transform-pair involves a symmetrical pair of Dirac functions as the frequency function. Thus, the time function

$$G(t) = \cos \omega_0 t \qquad\qquad (\omega_0 = 2\pi f_0)$$

corresponds to the frequency function

$$S(f) = \tfrac{1}{2}[\delta(f + f_0) + \delta(f - f_0)].$$

If the reader is disturbed over the fact that we are evidently going to base our analysis, at least initially, on the use of Dirac functions, he should note that Dirac functions are always paired with functions which are used widely and freely in transmission theory although they are not realistic in a physical sense. Functions of time, such as $\cos \omega_0 t$, which represent an infinitely long past and future history of activity, are not a bit more realistic in a physical sense than are "infinitely sharp" lines in the frequency spectrum. Similarly, functions of frequency, such as $\exp(-i\omega t_0)$, whose absolute values do not vanish as $f \to \infty$, are not a bit more realistic than impulsive "functions" of time. Nevertheless, as we

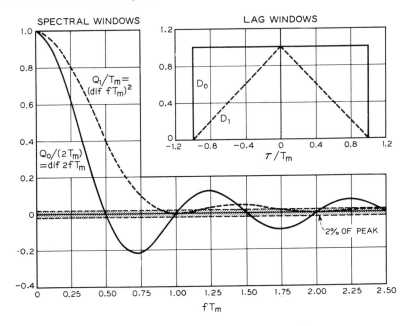

Fig. 14 — Lag windows D_0 and D_1. Spectral windows Q_0 and Q_1.

will see later on, these unrealistic pairs may be used as convenient bases for a wide variety of realistic pairs. They thus serve a very useful purpose.

The fifth transform-pair involves a *finite Dirac comb*tion, viz.

$$\ldots \cot \frac{-v}{2} \cdot \sin m\omega\Delta t = 2(m\cdot\Delta t)\cos(\pi f\cdot\Delta t)\frac{\operatorname{dif} 2f(m\cdot\Delta t)}{\operatorname{dif} f\cdot\Delta t}.$$

Except for a scale factor, the initial behaviour of this frequency function is illustrated in Fig. 9. Clearly, since $\cos 0 = \operatorname{dif} 0 = 1$, the limit of $Q_0(f; \Delta t)$, when $\Delta t \to 0$ with $m\cdot\Delta t = T_m$ held constant, is $Q_0(f)$. This corresponds to the formal convergence of $\nabla_m(t; \Delta t)$ to $D_0(t)$.

We have defined this finite Dirac comb with a half-sized Dirac function at each end because the corresponding frequency function has smaller side lobes, relative to the main lobe, than for the finite Dirac comb with a whole Dirac function at each end. This is easily seen from the fact that the effect of adding a further half-sized Dirac function at each end of $\nabla_m(t; \Delta t)$ is to add $\Delta t\cdot\cos m\omega\Delta t$ to $Q_0(f; \Delta t)$.

The frequency function $Q_0(f; \Delta t)$ is periodic, with a period of $1/\Delta t$ cps. It is symmetrical about every integral multiple of $1/(2\Delta t)$ cps. Thus, it has an absolutely maximum value of $2m\cdot\Delta t$ at the integral multiples of $1/\Delta t$ cps. It is zero at the integral multiples of $1/(2m\Delta t)$ cps which are not integral multiples of $1/\Delta t$ cps. For large values of m and small values of $\omega\Delta t$, it behaves approximately like $Q_0(f)$.

The sixth transform-pair involves an *infinite Dirac comb* in time, and, as it turns out, also an infinite Dirac comb in frequency. The time function is the formal limit of $\nabla_m(t; \Delta t)$ as $m \to \infty$, namely,

$$\nabla(t; \Delta t) = \Delta t \cdot \sum_{q=-\infty}^{q=\infty} \delta(t - q\Delta t).$$

The corresponding frequency function is

$$A\left(f; \frac{1}{\Delta t}\right) = \sum_{q=-\infty}^{q=\infty} \delta\left(f - \frac{q}{\Delta t}\right) = \Delta t\cdot\nabla\left(f; \frac{1}{\Delta t}\right).$$

This may be surmised from the fact that

$$\int_{-1/(2\Delta t)}^{1/(2\Delta t)} Q_0(f; \Delta t) \, df = 1 \qquad \text{for any } m$$

while

$$\lim_{m\to\infty} \int_{-\epsilon}^{\epsilon} Q_0(f; \Delta t) \, df = \frac{2}{\pi} \lim_{m\to\infty} Si(2\pi m\epsilon\Delta t), \quad \left(\text{where } Si(x) \equiv \int_0^x \frac{\sin y}{y} \, dy\right)$$

$$= 1 \text{ for any } \epsilon \text{ in } 0 < \epsilon < \frac{1}{2\Delta t}.$$

The result may indeed be obtained by applying the fourth transform-pair to the formal Fourier series representation of the infinite comb

$$\nabla(t; \Delta t) = 1 + 2 \sum_{q=1}^{q=\infty} \cos \frac{2\pi q t}{\Delta t}.$$

Since, with $T_m = m \cdot \Delta t$,

$$\nabla_m(t; \Delta t) = D_0(t) \cdot \nabla(t; \Delta t)$$

we also have, as we shall see in the next section,

$$Q_0(f; \Delta t) = Q_0(f) * A\left(f; \frac{1}{\Delta t}\right)$$

$$= \sum_{q=-\infty}^{\infty} Q_0\left(f - \frac{q}{\Delta t}\right).$$

The seventh transform-pair arises from the sixth by dividing by Δt on both sides.

A.3 *Convolution*

If $G(t) = G_1(t) \cdot G_2(t)$, then the Fourier transform of $G(t)$ may be expressed in terms of those of $G_1(t)$ and $G_2(t)$ as follows.

$$S(f) = \int_{-\infty}^{\infty} G_1(t) \cdot G_2(t) \cdot e^{-i\omega t} \, dt,$$

$$= \int_{-\infty}^{\infty} G_1(t) \cdot \left[\int_{-\infty}^{\infty} S_2(\xi) \cdot e^{i2\pi\xi t} \, d\xi\right] \cdot e^{-i\omega t} \, dt,$$

$$= \int_{-\infty}^{\infty} \left[\int_{-\infty}^{\infty} G_1(t) \cdot e^{-i2\pi(f-\xi)t} \, dt\right] \cdot S_2(\xi) \, d\xi,$$

$$= \int_{-\infty}^{\infty} S_1(f - \xi) \cdot S_2(\xi) \, d\xi.$$

This relation, in which $S_1(f)$ and $S_2(f)$ are interchangeable, is commonly expressed in the symbolic form

$$S(f) = S_1(f) * S_2(f)$$

The implied ~~

~~.....~~titute an *operational trans-*

(Convolution is often called by a variety of names such as Superposition theorem, Faltungsintegral, Green's theorem, Duhamel's theorem, Borel's theorem, and Boltzmann-Hopkinson theorem.)

It may be noted in the detailed derivation above (putting $f = 0$), that

$$\int_{-\infty}^{\infty} G_1(t) \cdot G_2(t) \, dt = \int_{-\infty}^{\infty} S_1{}^*(f) \cdot S_2(f) \cdot df$$

where $S_1{}^*(f)$ is the complex conjugate of $S_1(f)$. This is *Parseval's theorem* of which a very useful special case is

$$\int_{-\infty}^{\infty} [G(t)]^2 \, dt = \int_{-\infty}^{\infty} |S(f)|^2 \, df.$$

An example of convolution is supplied by the symmetrical triangular time function in the second transform-pair. This time function is the convolution of two symmetrical rectangular time functions from the first transform-pair, with appropriate scalar adjustments. Another example is the infinite Dirac comb $\nabla(t; \Delta t)$, which may be regarded as the convolution of the finite Dirac comb $\nabla_m(t; \Delta t)$ with the infinite Dirac comb $A(t; 2m\Delta t)$, that is

$$\nabla(t; \Delta t) = \nabla_m(t; \Delta t) * A(t; 2m\Delta t).$$

As the reader may easily verify, this corresponds to

$$A\left(f; \frac{1}{\Delta t}\right) = Q_0(f; \Delta t) \cdot \nabla\left(f; \frac{1}{2m\Delta t}\right).$$

Convolution of time functions occurs in communications systems whenever a signal is transmitted through a fixed linear network. If the input signal is $G_1(t)$, and if the impulse response of the network is $W(t)$, then the output signal is*

$$G(t) = \int_{-\infty}^{\infty} W(t - \lambda) \cdot G_1(\lambda) \, d\lambda$$

$$= W(t) * G_1(t).$$

The so-called *linear distortion* of the signal due to transmission through the network can be (and occasionally is) examined in terms of the effects of convolution, but the common practice among circuit engineers is to conduct the examination in terms of the corresponding frequency functions. There are good reasons for this common practice. The most important of these reasons are:

1. The relation between the frequency functions is simpler, viz.

$$S(f) = Y(f) \cdot S_1(f)$$

where $Y(f)$ is the transfer function of the network.

2. The effects of *amplitude distortion* of the signal and of *phase distortion* (of the unmodulated signal) may be examined independently. While phase distortion is critical in the transmission of pictures (facsimile), it is relatively unimportant in the transmission of speech or music.

3. The transmission characteristics of fixed linear networks are most easily calculated or measured accurately in terms of frequency rather than time.

4. Fixed linear network design techniques based on frequency functions are today much further developed (simpler, more powerful, and more versatile) than those based on time functions.

Convolution of frequency functions occurs in communications systems whenever a carrier wave is amplitude-modulated by a signal. If the input signal is $G_1(t)$, and if the carrier wave is $\cos \omega_0 t$, then the output signal, with *suppressed carrier*, is

$$G(t) = G_1(t) \cdot \cos \omega_0 t$$

---

* It may be of some help here to think of λ as "excitation time", and of t as "response time". In the equivalent formulation

$$G(t) = \int_{-\infty}^{\infty} W(\tau) \cdot G_1(t - \tau) \, d\tau$$

we may think of $\tau = t - \lambda$ as the "age" of input data at response time.

At this point attention is called to a device which will be used many times to simplify analysis, which is to use $-\infty$ and $+\infty$ as limits of integration, letting the integrand take care of the effective range of integration. In this case, if $G_1(\lambda) \equiv 0$ for $\lambda < t_0$, and $W(\tau) \equiv 0$ for $\tau < 0$, the effective range of integration would be $t_0 < \lambda < t$ or $0 < \tau < t - t_0$.

and the relation among the corresponding frequency functions is

$$S(f) = S_1(f) * \tfrac{1}{2} [\delta(f + f_0) + \delta(f - f_0)]$$

$$= 1 \quad \ \ \ $$

...... scheme under con-
...... sideband, single sideband, vestigial sideband, or
two-phase (as in TV chrominance signals). Further, the two-sided specifi-
cation of the modulated-carrier spectrum is essential for a correct picture
of the demodulation process used to recover the signal.

For present purposes we will be interested in convolution not only as
a tool for the synthesis of new transform-pairs but also as an analytical
tool. For example, by regarding a time function $G(t)$ as the product of
two other time functions $G_1(t)$ and $G_2(t)$ we can make use of the relation $S(f) = S_1(f) * S_2(f)$ to reach insights about $S(f)$ which do not
come easily from the explicit form of $S(f)$.

To make convolution a useful analytical tool, we have to visualize it
in some convenient way. This may be done in three ways. The relative
merits of these three points of view depend upon the circumstances in
any particular case.

In the first place, convolution may be visualized as a stretching process.
For example, in the equation

$$G(t) = \int_{-\infty}^{\infty} G_1(t - \lambda) \cdot G_2(\lambda) \, d\lambda$$

we visualize $G_2(\lambda) \cdot d\lambda$ as a rectangular element of $G_2(t)$, originally concentrated at $t = \lambda$. This rectangular element is then stretched into the
area under the elementary curve $G_1(t - \lambda) \cdot G_2(\lambda) \cdot d\lambda$ regarded as a function of t. This elementary curve has the shape of $G_1(t)$ with origin shifted
to $t = \lambda$. The total effect at any particular value of t is then obtained
by integration over λ. In this example, we have regarded $G_1(t)$ as the
"stretcher" operating on each element of $G_2(t)$. Of course, since convolution is commutative, we may interchange the roles of the two functions.

In the second place, if one of the functions in the convolution consists
exclusively of Dirac functions, each Dirac function may be regarded as
a "shifter" operating on the other function in the convolution. For ex-

ample,

$$\delta(t - a) * G(t) = \int_{-\infty}^{\infty} \delta(t - a - \lambda) \cdot G(\lambda) \, d\lambda = G(t - a).$$

In the third place, convolution may be visualized as a weighted integration with a moving weight function. For example, in the equation

$$G(t) = \int_{-\infty}^{\infty} G_1(t - \lambda) \cdot G_2(\lambda) \, d\lambda$$

we regard $G(t)$ as the integral of $G_2(\lambda)$ with weight function $G_1(t - \lambda)$. The position of the weight function with respect to the λ scale depends upon the value of t. In the event that the weight function has unit area, $G(t)$ may be regarded as the moving weighted average of $G_2(\lambda)$. (As previously noted, the roles of the two functions may be interchanged.)

As an example of the use of the ideas described above, let us assume that we have a function $G_0(t)$ which is zero outside of the interval $0 < t < T$, and for which the frequency function is $S_0(f)$. Let us generate a periodic function $G(t)$ by convolving $G_0(t)$ with $A(t; T)$. Then, since

$$G(t) = G_0(t) * A(t; T)$$

the frequency function corresponding to $G(t)$ is, from Item 7 of Table IV,

$$S(f) = S_0(f) \cdot \nabla \left(f; \frac{1}{T}\right).$$

As we expect, $S(f)$ consists of "lines" (of infinite height but finite area) at uniform intervals of $1/T$ cps. The complex intensities (areas) of these lines represent the amplitudes and relative phases of the terms in the conventional Fourier series representation of $G(t)$. Thus,

$$G(t) = \int_{-\infty}^{\infty} S(f) \cdot e^{i\omega t} \, df$$

$$= \frac{1}{T} \sum_{q=-\infty}^{q=\infty} S_0 \left(\frac{q}{T}\right) \cdot e^{(i2\pi q t)/T}.$$

As a second example, which is in a sense the dual of the first, let us assume that we have a function $G_0(t)$ for which the frequency function $S_0(f)$ is zero outside of the band $-f_0 < f < f_0$. Let us generate a discrete time series $G(t)$ by sampling $G_0(t)$ at uniform intervals of $1/(2f_0)$ seconds. If we regard sampling as a multiplication by (or as amplitude-modulation of) an infinite Dirac comb, then

$$G(t) = G_0(t) \cdot A\left(t; \frac{1}{2f}\right).$$

Hence, tʰ

... ...quency function $S_1(f)$,

$$S_1(f) = \frac{1}{2f_0}, \quad |f| < f_0$$

$$= 0, \quad |f| > f_0$$

it will revert to $S_0(f)$. Thus,

$$S_1(f) \cdot S(f) = S_0(f).$$

Hence, if $G_1(t)$ is the time function corresponding to $S_1(f)$, namely,

$$G_1(t) = \frac{\sin \omega_0 t}{\omega_0 t},$$

then

$$G_1(t) * G(t) = G_0(t).$$

Thus, sampling $G_0(t)$ to get the discrete time series $G(t)$, and convolving $G(t)$ with $G_1(t)$, restores $G_0(t)$ exactly. This result reflects the well-known *sampling theorem* in information theory. The effect of sampling $G_0(t)$ at uniform intervals of other than $1/(2f_0)$ seconds is readily visualized.

A.4 *Windows*

If a time function is even (and of course real), the corresponding frequency function is real (and of course even), and conversely. These circumstances will prevail when we deal with autocovariance functions, power spectra, and appropriate weight functions. Under these circumstances, the weight functions will be called *windows*. Such windows will be considered in transform-pairs, and the members of any pair will be distinguished as the *lag window*, and the *spectral window*.

Time windows convolved with periodic functions of time have been used by Guillemin,[22] under the name "scanning functions", to examine the behavior of weighted partial sums of Fourier series. We use them in Sections B.4 and B.10 where we call them *data windows*, and their Fourier transforms (which may be complex) *frequency windows*.

A.5 *Realistic Pairs from Unrealistic Pairs*

Transform-pairs which involve Dirac functions are very easily converted into a wide variety of realistic pairs. As an example, let us consider the seventh pair (infinite Dirac combs) which requires two convolutions for conversion to a realistic pair. If we convolve the time functions of the first and seventh pairs, taking $T_m \ll \Delta t$, we get a time function which represents an infinite train of narrow rectangular pulses of unit height. The corresponding frequency function still consists of Dirac functions but these now do not have a uniform intensity. If we next multiply the time function of this pair by the time function of the first pair, taking $T_m \gg \Delta t$, we get a time function which represents a long but finite train of narrow rectangular pulses. The corresponding frequency function is continuous and consists chiefly of very narrow peaks of finite height approaching zero as $f \to \infty$.

A sinusoidal carrier wave of finite though great length may be represented as the product of the time functions of the first and fourth pairs with $T_m \gg 1/f_0$. The corresponding frequency function is continuous and consists of very narrow peaks at $\pm f_0$, with much lower subsidiary peaks of height approaching zero as $f \to \infty$.

If the time function of the third pair is convolved with the time function

$$
\begin{aligned}
G(t) &= 0 & t &< 0 \\
&= \frac{1}{T}\, e^{-t/T} & t &> 0 ,
\end{aligned}
$$

the resultant frequency function is

$$
S(f) = \frac{1}{1 + i\omega T}\, e^{-i\omega t_0}
$$

of which the absolute value falls off asymptotically like $1/f$ as $f \to \infty$, however small $T(>0)$ might be.

In line with this discussion, it should be noted that a realistic "white noise" spectrum must be effectively band-limited by an asymptotic fall-off at least as fast as $1/f^2$. Under certain circumstances, however, we may assume that the spectrum is flat to any frequency. Let us suppose that the spectrum is in fact

$$P(f) = \frac{\dfrac{\sigma^2}{\pi f_c}}{}$$

$$P(f) \approx \frac{\sigma}{\pi f_c}$$

and, therefore, that

$$C(\tau) \approx \frac{\sigma^2}{\pi f_c} \delta(\tau)$$

although such an assumption is unrealistic if carried to indefinitely high frequencies (the input noise would have infinite variance). Hence, if the impulse response of the network is $W(t)$, the autocovariance of the output noise is

$$
\begin{aligned}
C_{\text{out}}(t_i - t_j) &= \text{ave}\left\{ \int_{-\infty}^{\infty} W(\tau_1) X(t_i - \tau_1)\, d\tau_1 \right.\\
&\qquad \left. \int_{-\infty}^{\infty} W(\tau_2) \cdot X(t_j - \tau_2)\, d\tau_2 \right\}\\
&= \int\!\!\!\int_{-\infty}^{\infty} W(\tau_1) W(\tau_2) \cdot C(t_i - t_j - \tau_1 + \tau_2) d\tau_1\, d\tau_2\\
&\approx \frac{\sigma^2}{\pi f_c} \int_{-\infty}^{\infty} W(\tau_1) \cdot W(\tau_1 - t_i + t_j)\, d\tau_1 .
\end{aligned}
$$

In particular, the variance of the output noise is

$$C_{\text{out}}(0) \approx \frac{\sigma^2}{\pi f_c} \int_{-\infty}^{\infty} [W(\tau_1)]^2\, d\tau_1$$

which by Parseval's theorem is equivalent to

$$C_{\text{out}}(0) \approx \frac{\sigma^2}{\pi f_c} \int_{-\infty}^{\infty} |Y(f)|^2\, df$$

where $Y(f)$ is the transfer function of the network. These results are realistic. (The variance of the output noise is finite and approximately correct).

A.6 *Some Trigonometric Identities*

In this section we develop some trigonometric identities which will be needed later on. We start with the equation

$$\sum_{-a}^{b} \cos (\psi + 2hu) = \frac{\sin (a + b + 1)u}{\sin u} \cos [\psi + (b - a)u]$$

which is easily obtained by substituting

$$\cos x = \frac{e^{ix} + e^{-ix}}{2}$$

in the left-hand member, summing the exponential terms and making some elementary trigonometric substitutions. By substituting $\psi \pm \pi/2$ for ψ we then get

$$\sum_{-a}^{b} \sin (\psi + 2hu) = \frac{\sin (a + b + 1)u}{\sin u} \sin [\psi + (b - a)u].$$

Now, setting $u = \pi f$, and using the function introduced in Section A.2,

$$\frac{\sin pu}{p \sin u} = \frac{(\sin p\pi f)/p\pi f}{(\sin \pi f)/\pi f} = \frac{\text{dif } pf}{\text{dif } f},$$

which, on differentiation, yields

$$\frac{d}{df} \frac{(\text{dif } pf)}{(\text{dif } f)} = \left(\frac{\text{dif } pf}{\text{dif } f}\right) \left(p \frac{\text{dif}' pf}{\text{dif } pf} - \frac{\text{dif}' f}{\text{dif } f}\right).$$

Before we rewrite our summation formulas in terms of such ratios of "dif" functions, we need to appreciate their behavior. For p not very small, (dif pf)/(dif f) behaves much like the numerator for pf small and moderate. The effect of the denominator is to force symmetry around integer multiples of $\frac{1}{2}$, so that the peak at $f = 0$ is repeated at $f = 1, 2, 3, \cdots$, thus making its behavior consistent with aliasing. For $0 \leqq f \leqq \frac{1}{2}$ its other effects are minor, since in this range $(2/\pi) \leqq$ dif $f \leqq 1$, while the extrema of dif pf have shrunk from $+1$ to $\pm 2/(p\pi)$. For most considerations, therefore, we can approximate this ratio by the numerator.

We now rewrite our summations as means, introducing (dif pf)/(dif f), finding

$$\frac{1}{a + b + 1} \sum_{-a}^{b} \cos (\psi + 2h\pi f) = \frac{\text{dif } (a + b + 1)f}{\text{dif } f} \cos [\psi + (b - a)\pi f]$$

$$\frac{1}{a + b + 1} \sum_{-a}^{b} \sin (\psi + 2h\pi f) = \frac{\text{dif } (a + b + 1)f}{\text{dif } f} \sin [\psi + (b - a)\pi f].$$

Differentiating with respect to f, and multiplying through by

$$-(a + b + 1)/(2\pi),$$

we get

$$-\frac{a + b + 1}{2\pi} \frac{\text{dif}' f}{\text{dif} f}\Big) \cos[\psi + (b - a)\pi f]\Big]$$

with a similar formula for

$$\sum_{-a}^{b} h \cos(\psi + 2h\pi f).$$

We shall now use these formulas to obtain results about the average values of certain quadratic functions of chance variables X_0, X_1, $\cdots$, X_n. The average value of any such quadratic function can be represented in terms of a corresponding spectral window $Q(f)$ in the form

$$\int_0^\infty Q(f) \cdot 2P(f)\, df$$

whenever

$$\text{ave}\, \{X_t X_{t+q}\} = \int_0^\infty \cos 2\pi q f \cdot 2P(f)\, df$$

for all suitable integers t and q, since the quadratic function can be expressed as a sum of multiples of terms of the form $X_t X_{t+q}$. To determine the height, $Q(f_0)$, of the spectral window corresponding to a specific quadratic function, it suffices to consider the special case $2P(f) = \delta(f - f_0)$, for which ave $\{X_t X_{t+q}\} = \cos 2\pi q f_0$, when the average value of the quadratic function for such a special set of X_t is exactly $Q(f_0)$.

If ave $\{X_t X_{t+q}\} = \cos 2\pi q f$, we easily find that

$$\text{ave}\left\{\left(\frac{1}{a + b + 1} \sum_{-a}^{b} X_h\right)\left(\frac{1}{c + d + 1} \sum_{-c}^{d} X_g\right)\right\}$$

$$= \frac{1}{a + b + 1} \sum_{-a}^{b} \frac{1}{c + d + 1} \sum_{-c}^{d} \cos 2\pi f(g - h)$$

$$= \frac{1}{a+b+1} \sum_{-a}^{b} \frac{\mathrm{dif}\ (c+d+1)f}{\mathrm{dif}\ f} \cos\ (-2\pi f h + (d-c)\pi f)$$

$$= \frac{\mathrm{dif}\ (a+b+1)f}{\mathrm{dif}\ f} \frac{\mathrm{dif}\ (c+d+1)f}{\mathrm{dif}\ f} \cos\ (d-c-a+b)\pi f$$

$$\approx \mathrm{dif}\ (a+b+1)f\ \mathrm{dif}\ (c+d+1)f \cos\ (d-c-a+b)\pi f$$

any of these expressions being the spectral window corresponding to

$$\left(\frac{1}{a+b+1} \sum_{-a}^{b} X_h \right) \left(\frac{1}{c+d+1} \sum_{-c}^{d} X_g \right).$$

Making the same assumption, we find that (where $n = 2\ell + 1$)

$$\mathrm{ave}\ \left\{ \left(\sum_{-\ell}^{\ell} hX_h \right) \left(\sum_{-\ell}^{\ell} gX_g \right) \right\}$$

$$= \sum_{-\ell}^{\ell} \sum_{-\ell}^{\ell} gh\ \cos\ 2\pi f(g-h)$$

$$= \sum_{-\ell}^{\ell} g \left(\frac{\mathrm{dif}\ nf}{\mathrm{dif}\ f} \right) \left(\frac{n^2}{2\pi} \frac{\mathrm{dif}'\ nf}{\mathrm{dif}\ nf} - \frac{n}{2\pi} \frac{\mathrm{dif}'\ f}{\mathrm{dif}\ f} \right) \sin\ 2\pi fg$$

$$= \frac{n^4}{4} \left(\frac{\mathrm{dif}\ nf}{\mathrm{dif}\ f} \right)^2 \left(\frac{1}{\pi} \frac{\mathrm{dif}'\ nf}{\mathrm{dif}\ nf} - \frac{1}{n\pi} \frac{\mathrm{dif}'\ f}{\mathrm{dif}\ f} \right)^2.$$

These expressions therefore represent the spectral windows corresponding to

$$\left(\sum_{-\ell}^{\ell} hX_h \right)^2.$$

PART II

.... of this part will be numbered in exact correspondence
with the sections of the general account. Thus, for example, Section
B.7, below, presents the details and sidelights related to Section 7,
of Part I. (Certain sections will be omitted.)

B.1 GAUSSIAN PROCESSES AND MOMENTS

There are two common modes of description of a random process,
intuitively quite different. One uses the idea of an ensemble, the other a
function of infinite extent. The first is undoubtedly more flexible, as it
can describe processes, even non-stationary (e.g. evolving) ones, which
cannot be described by any single function, even one of infinite extent.
The first is also, at least in the eyes of the statistician, more fundamental, since uncertainty, which he regards as a central concept, enters
directly and explicitly. It is possible to regard the single-function approach as an attempt to minimize recognition of the statistical aspects
of the situation. Once, such minimization may have been of some value,
but today the essentially statistical nature of communication, be it of
symbols, voice, picture or feedback information, is well established. The
communication engineer is aware that he must have designed not only
for the message which was sent, but also for the one which might have
been sent — moreover that his design demanded consideration of the
relative probabilities of various messages that might have been sent
(and were not).

Such a statistical view of message or noise confronts us with the need,
not only of picking out what functions might arise, but also of attaching
probabilities to functions (at least to sets of functions). To do this directly and completely requires much careful mathematics. As far as
questions associated with observations and data are concerned, there is,

fortunately, no need for such care and complexity. We know that any empirical time function can be adequately represented by some finite number, large or small, of ordinates. Thus, for practical purposes, it suffices to be able to assign probabilities to sets of n ordinates — for n finite but possibly quite large. It is for this reason that we went directly to probability distributions of such n-dimensional sections in the general account.

This replacement of a continuous record by discrete ordinates is related to the sampling theorem of information theory. The relationship is, regrettably, not quite simple. Given a band-limited "signal" defined for all time from $-\infty$ to $+\infty$, and moderately well-behaved otherwise, the sampling theorem (Nyquist[23]), which is also known as the Cardinal Theorem of Interpolation Theory (Whittaker[24]), states that equi-spaced ordinates, if close enough together, extending from $-\infty$ to $+\infty$ will precisely determine the function. Given a band-limited function over a finite interval, the corresponding result is almost practically true. It is not true in a precise impractical sense, since every band-limited function can be obtained from an entire function of exponential type (of a complex variable) by considering only the values taken on along the real axis. Consequently, if we know a band-limited function precisely in an interval, its values are determined everywhere. Theoretically determined, but not practically so, since the kernels expressing this determination behave like hyperdirective antennas. Since the values at equi-spaced points in the interval do not determine the values at equi-spaced points outside the interval (which would be determined by precise knowledge throughout the interval) the latter cannot be obtained from the equi-spaced values in the interval. In practice, however, functions are not quite band-limited, and measurements always involve measurement noise. When these two facts are considered, a sufficiently closely spaced set of equi-spaced ordinates extracts all the practically useful information in the continuous record.

The fact that averages, variances and covariances completely characterize any n-dimensional Gaussian distribution is common statistical knowledge, and follows by inspection of the conventional general form of Gaussian probability density function, in which these moments appear as the only parameters.

B.2 AUTOCOVARIANCE FUNCTIONS AND POWER SPECTRA FOR PERFECT INFORMATION

If $X(t)$ is a function generated by a stationary Gaussian process, and if

$$\lim_{T \to \infty} \frac{1}{T} \int_{-T/2}^{T/2} X(t) \, dt = 0 \qquad \text{(B-2.1)}$$

then the *autocovariance function* of the process is

$$C(\tau) = \lim_{T \to \infty} \frac{1}{T} \int_{-T/2}^{T/2} X(t) \cdot X(t + \tau) \, dt. \tag{B-2.2}$$

In particular, the *variance* of the process is

$$C(0) = \lim_{T \to \infty} \frac{1}{T} \int_{-T/2}^{T/2} [X(t)]^2 \, dt. \tag{B-2.3}$$

If the function $X(t)$ is passed through a fixed linear network whose impulse response (response to a unit impulse applied at $t = 0$) is $W(t)$, then the output of the network will be

$$X_{\text{out}}(t) = \int_{-\infty}^{\infty} W(\lambda) \cdot X(t - \lambda) \cdot d\lambda \,,$$

and the autocovariance of the output will be

$$\begin{aligned} C_{\text{out}}(\tau) &= \lim_{T \to \infty} \frac{1}{T} \int_{-T/2}^{T/2} \int_{-\infty}^{\infty} \int_{-\infty}^{\infty} W(\lambda_1) \cdot W(\lambda_2) \cdot X(t - \lambda_1) \\ &\qquad\qquad \cdot X(t + \tau - \lambda_2) \cdot d\lambda_1 \cdot d\lambda_2 \cdot dt \\ &= \int_{-\infty}^{\infty} \int_{-\infty}^{\infty} W(\lambda_1) \cdot W(\lambda_2) \cdot C(\tau + \lambda_1 - \lambda_2) \cdot d\lambda_1 \cdot d\lambda_2. \end{aligned}$$

If we now let

$$C(\tau) = \int_{-\infty}^{\infty} P(f) e^{i\omega\tau} \, df \qquad (\omega = 2\pi f), \tag{B-2.4}$$

then

$$C_{\text{out}}(\tau) = \int_{-\infty}^{\infty} \int_{-\infty}^{\infty} \int_{-\infty}^{\infty} W(\lambda_1) \cdot W(\lambda_2) \cdot P(f) \cdot e^{i\omega(\tau + \lambda_1 - \lambda_2)} \cdot d\lambda_1 \cdot d\lambda_2 \cdot df.$$

But

$$\int_{-\infty}^{\infty} W(\lambda_2) \cdot e^{-i\omega\lambda_2} \cdot d\lambda_2 = Y(f)$$

is the transfer function (ratio of steady-state response to excitation, when the excitation is $e^{i\omega t}$) of the network, and

$$\int_{-\infty}^{\infty} W(\lambda_1) \cdot e^{i\omega\lambda_1} \cdot d\lambda_1 = Y(-f)$$

is the complex-conjugate of $Y(f)$. Hence,

$$C_{\text{out}}(\tau) = \int_{-\infty}^{\infty} | Y(f) |^2 \cdot P(f) \cdot e^{i\omega\tau} \cdot df. \tag{B-2.5}$$

In particular, the variance of the output is

$$C_{\text{out}}(0) = \int_{-\infty}^{\infty} |Y(f)|^2 \cdot P(f) \cdot df. \tag{B-2.6}$$

Since $|Y(f)|^2$ is the power transfer function of the network, it is natural to call $P(f)$ the *power spectrum* of the process. The power spectrum has the dimensions of variance per cycle per second. By equation (B-2.4) we have

$$P(f) = \int_{-\infty}^{\infty} C(\tau) \cdot e^{-i\omega\tau} \, d\tau, \quad (\omega = 2\pi f), \quad \text{(B-2.7)}$$

and, by equation (B-2.5),

$$P_{\text{out}}(f) = |Y(f)|^2 \cdot P(f).$$

Autocovariance functions and power spectra are usually regarded as one-sided functions of lag and frequency, respectively, related by the formulae (Rice[3] p. 285),

$$P(f) = 4 \int_{0}^{\infty} C(\tau) \cdot \cos \omega\tau \cdot d\tau, \qquad \text{(not used here),}$$

$$C(\tau) = \int_{0}^{\infty} P(f) \cdot \cos \omega\tau \cdot df, \qquad \text{(not used here).}$$

However, we will find it very convenient for analytical purposes, to continue to regard them as two-sided even functions related by equations (B-2.4) and (B-2.7). This will be evident in Section B.4 where spectral windows will be convolved with power spectra, and in Section B.6 where we would otherwise have to make use of rather complicated trigonometric identities.

In a few places where we contemplate computations, we may write

$$P(f) = 2 \int_{0}^{\infty} C(\tau) \cdot \cos \omega\tau \cdot d\tau, \qquad (\omega = 2\pi f),$$

instead of equation (B-2.7), but it is important to observe that the power spectrum $P(f)$ thus obtained must still be regarded as a two-sided even function which contains only a half of the total power or variance in the positive frequency range. If we prefer to think ultimately of a one-sided power spectrum (over only positive frequencies), in accordance with engineering practice, then we should take $2P(f)$ as the power density (per cycle per second) for positive frequencies.

The power spectrum $P(f)$ may also be expressed directly in terms of $X(t)$. The formal derivation of this expression on the basis of the formula (B-2.2) for the autocovariance function is complicated by the fact that the integral

$$\int_{-T/2}^{T/2} X(t) \cdot X(t + \tau) \, dt$$

actually depends upon two more-or-less distinct pieces of $X(t)$, one in the range $-T/2 < t < T/2$, the other in the range

$$-\frac{T}{2} - \tau < t < \frac{T}{2} - \tau.$$

We may avoid this complication by using the equivalent formula

$$C(\tau) = \lim_{T \to \infty} \frac{1}{T} \int_{-\infty}^{\infty} G(t) \cdot G(t + \tau) \cdot dt, \qquad \text{(B-2.8)}$$

where

$$G(t) = X(t), \qquad |t| < \frac{T}{2},$$

$$= 0, \qquad |t| > \frac{T}{2}.$$

Let

$$S(f) = \int_{-\infty}^{\infty} G(t) \cdot e^{-i\omega t} \, dt,$$

so that

$$G(t) = \int_{-\infty}^{\infty} S(f) \cdot e^{i\omega t} \, df.$$

Then

$$C(\tau) = \lim_{T \to \infty} \frac{1}{T} \int_{-\infty}^{\infty} G(t) \cdot \int_{-\infty}^{\infty} S(f) \cdot e^{i\omega(t+\tau)} \, df \cdot dt$$

$$= \lim_{T \to \infty} \frac{1}{T} \int_{-\infty}^{\infty} S(f) \cdot e^{i\omega\tau} \int_{-\infty}^{\infty} G(t) \cdot e^{i\omega t} \, dt \cdot df$$

$$= \lim_{T \to \infty} \frac{1}{T} \int_{-\infty}^{\infty} S(f) \cdot S(-f) \cdot e^{i\omega\tau} \, df.$$

Comparing with (B-2.4) we get, at least formally,

$$P(f) = \lim_{T \to \infty} \frac{1}{T} \mid S(f) \mid^2$$

$$= \lim_{T \to \infty} \frac{1}{T} \left| \int_{-\infty}^{\infty} G(t) \cdot e^{-i\omega t} \, dt \right|^2 \qquad \text{(B-2.9)}$$

$$= \lim_{T \to \infty} \frac{1}{T} \left| \int_{-T/2}^{T/2} X(t) \cdot e^{-i\omega t} \, dt \right|^2.$$

(See Rice,[3] p. 320, and Bennett,[25] p. 621.)

B.3 THE PRACTICAL SITUATION

In the study of second moments of random processes, the balance between the approach through autocovariances and the approach through power spectra is, in at least one sense, a little closer than Section 3 would seem to imply. As a means for understanding, and as a guide for intelligent design, the power spectrum is without a peer. The autocovariance function is of little use except as a basis for estimating the power spectrum. This is fundamentally because, in most physical systems, power spectra have reasonable shapes, are relatively easily understandable, and often are quite directly influenced by the basic variables of the situation, whatever these may be. The process of using an empirically observed and analyzed power spectrum usually goes through some such chain of steps as this:

(i) Planning and design.

(ii) Observation and recording.

(iii) Analysis and preparation of estimates.

(iv) Comparison of estimates with existing and synthesizable theoretical structures and quantitative information.

(v) Selection of the best working version of a theoretically-guided approximation to the estimates.

(vi) Use of this working version.

Our theoretical understanding of the situation, and of the forces in it, play important roles, which we should never allow ourselves to forget, in steps (i) and (v). (We use this understanding to the utmost, in its proper place, but we do not, and should not, allow it to narrow down steps (ii) and (iii) to the point where we have little or no chance of discovering that it was incomplete or in error. Thus, we estimate a considerable number of smoothed spectral densities, and not merely a few constants of a suggested theoretical curve. After we have compared curve and points, we may then wish to estimate the constants.)

The simplest and most straightforward use we can make of the power spectrum is to predict the output spectrum, or perhaps only the output power, when the process studied provides the input to a linear device with a known power transfer function.

Except for purely descriptive uses, checking on performance for agreement with anticipation, and for predicting the behavior of already designed linear systems, the most elementary use of a power spectrum lies in optimizing the performance of some linear predictor or filter as measured á la least squares. The nature of this situation is not quite that one which most persons imagine.

If we really have no theoretical insight into the situation *at all*, we might as well (nay, perhaps, might better) stay in the time domain. We have autocovariances (say) for some limited range of lags. If the duration of the transient response of our filter or predictor is not going to exceed one-half this time limit, then we can write out the estimated variance of any predictor or filter directly in terms of our estimates of autocovariances and of the time description of the filter or predictor, and could then minimize this directly. With no theoretical insight this should work at least as well as any other way. With no theoretical insight, analysis in the time domain would be relatively good, perhaps even optimal — and probably absolutely poor.

But we do not, and almost always should not, optimize filters or predictors in this way. The reason is simple. Actual power spectra are often simple and understandable. Actual autocovariance functions are hardly, if ever, simple and understandable. The intervention of theoretical insight and human judgment at step (v) is crucial and valuable. This intervention is effective in the frequency domain, but not in the time domain. (Step (v) is likely to stand for some time, as a challenge to the ability of statisticians to wisely and effectively automatize inferential procedures.)

An additional advantage of power spectrum analysis over autocovariance analysis was pointed out in Section 3, namely the ease of compensation for (linear) modification before measurement. When the random function $X(t)$ is passed through a time-invariant linear transmission system whose impulse response is $W(t)$, the output random function, which may be the only function accessible for measurement, is

$$X_{\text{out}}(t) = \int_{-\infty}^{\infty} W(\tau_1) \cdot X(t - \tau_1) \, d\tau_1.$$

The relation between the autocovariance $C_{\text{out}}(\tau)$ of the modified process, and the autocovariance $C(\tau)$ of the original process may be derived as follows:

$$C_{\text{out}}(\tau) = \text{ave } \{X_{\text{out}}(t) \cdot X_{\text{out}}(t + \tau)\}$$

$$= \text{ave } \left\{ \iint\limits_{-\infty}^{\infty} W(\tau_1) \cdot X(t - \tau_1) \cdot X(t + \tau - \tau_2) \cdot W(\tau_2) \, d\tau_1 \, d\tau_2 \right\}$$

$$= \iint\limits_{-\infty}^{\infty} W(\tau_1) \cdot C(\tau + \tau_1 - \tau_2) \cdot W(\tau_2) \, d\tau_1 \, d\tau_2.$$

Putting $\tau_2 = \tau_1 + \lambda$, we get

$$C_{\text{out}}(\tau) = \int_{-\infty}^{\infty} C(\tau - \lambda) \cdot \left[\int_{-\infty}^{\infty} W(\tau_1) \cdot W(\tau_1 + \lambda) \, d\tau_1 \right] d\lambda$$

$$= C(\tau) * W(\tau) * W(-\tau).$$

Measurement of $X_{\text{out}}(t)$ can give estimates of $C_{\text{out}}(\tau)$ which must subsequently be converted into estimates of $C(\tau)$. The only practical way to make this conversion seems to be through Fourier transformation of the estimates of $C_{\text{out}}(\tau)$ into the frequency domain, compensation there, and Fourier retransformation. Such a procedure, in effect, invokes the relation between the power spectrum $P_{\text{out}}(f)$ of the modified process and the power spectrum $P(f)$ of the original process. This relation is

$$P_{\text{out}}(f) = P(f) \cdot |Y(f)|^2,$$

where $Y(f)$ is the transfer function corresponding to the impulse response $W(t)$.

Details for Continuous Analysis

B.4 Power Spectrum Estimation from a Continuous Record of Finite Length

In the ideal case considered in Section B.2, which assumes that we have an infinite length of $X(t)$, we can calculate the power spectrum $P(f)$ in two ways — either directly from $X(t)$, or indirectly as the Fourier transform of the autocovariance function $C(\tau)$ which is calculable directly from $X(t)$. The basic choice is, leaving limiting problems aside, between squaring a Fourier transform, or Fourier transforming an average of products. In either case multiplication and Fourier transformation must enter.

From the point of view of the ensemble, as opposed to the single time function, we seek to estimate a particular basis for the second moments of all linear combinations. We may estimate any convenient basis as a start, and then transform.

Clearly, from either point of view, any result obtained in one way can also be obtained in the other. Differences between a time approach and a frequency approach must be differences in (i) ease of understanding, (ii) ease of manipulation of formulas, (iii) ease of calculation with numbers, rather than in anything more essential.

Understanding and a simple description of the procedure which yields reasonably stable estimates, and which we have discussed in general, is more easily obtained by the indirect route, so we shall proceed accordingly, beginning with a general outline of a hypothetical procedure for power spectrum estimation from a continuous record of finite length.

A general outline of a hypothetical procedure for power spectrum estimation from a continuous record of finite length, specifically $X(t)$ for $-T_n/2 \leq t \leq T_n/2$, is as follows:

(1) *Calculate the apparent autocovariance function*

$$C_{00}(\tau) = \frac{1}{T_n - |\tau|} \int_{-(T_n-|\tau|)/2}^{(T_n-|\tau|)/2} X\left(t - \frac{\tau}{2}\right) \cdot X\left(t + \frac{\tau}{2}\right) dt \quad \text{(B-4.1)}$$

for $|\tau| \leq T_m < T_n$, where T_n is the length of the record, and T_m is the maximum lag to be used. We shall see in Section B.9 that the stability of our power spectrum estimates depends upon how small we take the ratio T_m/T_n.

(For the purpose of the theoretical analysis in this section we assume that the data contain no errors of measurement; in particular, no bias due to a displaced (perhaps drifting) zero. The effects of such errors are considered elsewhere.)

(2) *Calculate the modified apparent autocovariance function*

$$C_i(\tau) = D_i(\tau) \cdot C_{00}(\tau), \quad \text{(B-4.2)}$$

where $D_i(\tau)$ is a prescribed lag window, an even function such that

$$D_i(0) = 1,$$

and

$$D_i(\tau) = 0 \quad \text{for} \quad |\tau| > T_m.$$

Note that $C_i(\tau) = 0$ for $|\tau| > T_m$ although $C_{00}(\tau)$ is not available for $|\tau| > T_m$.

(3) *Calculate the estimated power spectrum*

$$P_i(f) = 2 \int_0^\infty C_i(\tau) \cdot \cos \omega\tau \cdot d\tau. \quad \text{(B-4.3)}$$

Our object is now to determine the relation between ave $\{P_i(f)\}$ and $P(f)$, where "ave" denotes the ensemble average, that is, the average over all possible continuous pieces of $X(t)$ of length T_n. (The variability (specifically, the variance) of $P_i(f)$ will be examined in Section B.9.) Since $C_{00}(\tau)$ is not calculated for $|\tau| > T_m$, it is clear that

$$\text{ave } \{C_{00}(\tau)\} = C(\tau), \qquad \text{only for } |\tau| \leq T_m.$$

However, because $D_i(\tau) = 0$ for $|\tau| > T_m$,

$$\text{ave } \{C_i(\tau)\} = D_i(\tau) \cdot C(\tau), \qquad \text{for any } \tau.$$

Hence,

$$\text{ave}\{P_i(f)\} = \int_{-\infty}^{\infty} D_i(\tau) \cdot C(\tau) \, e^{-i\omega\tau} \, d\tau.$$

Then, if $Q_i(f)$ is the Fourier transform of $D_i(\tau)$, the relation we seek is, symbolically,

$$\text{ave } \{P_i(f)\} = Q_i(f) * P(f),$$

or explicitly,

$$\text{ave } \{P_i(f_1)\} = \int_{-\infty}^{\infty} Q_i(f_1 - f) \cdot P(f) \, df. \qquad \text{(B-4.4)}$$

This relation is in a form, (B-2.6), which is familiar to communications engineers except for the fact that $Q_i(f_1 - f)$ is not an even function of f, when $f_1 \neq 0$, and may be negative in some ranges of f. However, taking advantage of the fact that $P(f)$ is an even function of f, we may write

$$\text{ave}\{P_i(f_1)\} = \int_0^{\infty} H_i(f; f_1) \cdot P(f) \, df, \qquad \text{(B-4.5)}$$

where

$$H_i(f; f_1) = Q_i(f + f_1) + Q_i(f - f_1). \qquad \text{(B-4.6)}$$

The function $H_i(f; f_1)$ is an even function of f as well as of f_1. Hence it satisfies one of the necessary conditions for a physically realizable *power transfer function*. However, inasmuch as it may be negative in some ranges of f (actually an advantage as we will see), it may still not be physically realizable. Nevertheless, it is convenient to regard $H_i(f; f_1)$ as the power transfer function of a network, and to regard ave $\{P_i(f_1)\}$ as the long-time-average power output of the network when continuously driven by the random process.

Since $P_i(f_1)$, whether calculated from a single piece of $X(t)$ as outlined above, or calculated as an average over a finite number of pieces of $X(t)$, is an estimate of ave $\{P_i(f_1)\}$ in the usual statistical sense, it is evident that the calculated power density $P_i(f_1)$ is an estimate of an average-over-frequency of the true power spectrum $P(f)$, and not an estimate of the local power density $P(f_1)$. The calculated power density $P_i(f_1)$ may be regarded as an estimate of the local power density $P(f_1)$ only to the degree to which $Q_i(f)$ approximates $\delta(f)$. However, under the restriction that $D_i(\tau) = 0$ for $|\tau| > T_m$, the degree to which $Q_i(f)$ approximates $\delta(f)$ depends chiefly on how large we take T_m. On the other hand, as we will find in Section B.9, the larger we take T_m/T_n the less stable will the estimates be. Hence, in general, it will be wasteful to demand more frequency resolution than we actually need. In many cases we may even have to take less frequency resolution than we would like to have, in order to secure a reasonable stability of the estimates. Clearly, for any specific value of T_n, (and number of pieces of record), we can increase frequency resolution (or stability) only by sacrificing stability (or frequency resolution).

We have just examined a hypothetical method of power spectrum estimation, in which we compute an apparent autocovariance function, modify it, and take the cosine transform. We will now examine a method, also hypothetical, in which we modify the data, take the sine and cosine transforms, compute the sum of the squares at each frequency, and divide by the length of the record. If the data is $X(t)$ for $0 < t < T_n$, and the weighting function (data window) is $B_i(t)$, the estimated power spectrum is computed essentially according to the formula

$$P_{ei}(f) = \frac{1}{T_n} \left| \int_0^{T_n} B_i(t) \cdot X(t) \cdot e^{-i\omega t} \, dt \right|^2. \qquad \text{(B-4.7)}$$

To determine the average it is convenient to assume that $X(t)$ is of unlimited extent, to specify $B_i(t)$ to be identically zero for $t < 0$ and $t > T_n$, and to allow the *data window* $B_i(t)$ to be located anywhere in time by substituting $B_i(t - \lambda)$ for $B_i(t)$. Then

$$P_{ei}(f;\lambda) = \frac{1}{T_n} \left| \int_{-\infty}^{\infty} B_i(t - \lambda) \cdot X(t) \cdot e^{-i\omega t} \, dt \right|^2. \qquad \text{(B-4.8)}$$

This is the Fourier transform of

$$C_{ei}(\tau;\lambda) = \frac{1}{T_n} \int_{-\infty}^{\infty} B_i(t - \lambda) \cdot X(t) \cdot B_i(t - \lambda + \tau) \cdot X(t + \tau) \cdot dt. \qquad \text{(B-4.9)}$$

[Compare with (B-2.8) and (B-2.9)]. Now, since the random process is stationary, the ensemble average is equivalent to the average over λ; that is,

$$\text{ave } \{C_{ei}(\tau)\} = \lim_{T\to\infty} \frac{1}{T} \int_{-T/2}^{T/2} C_{ei}(\tau;\lambda) \cdot d\lambda.$$

Substituting (B-4.9) and $\lambda = t - \xi$ into the right-hand member, we get

$$\text{ave } \{C_{ei}(\tau)\} = \frac{1}{T_n} \cdot \lim_{T\to\infty} \frac{1}{T} \int_{-\infty}^{\infty} X(t) \cdot X(t+\tau)$$
$$\cdot \left[\int_{t-(T/2)}^{t+(T/2)} B_i(\xi) \cdot B_i(\xi+\tau) \cdot d\xi \right] \cdot dt.$$

Reversing the order of integration we get

$$\text{ave } \{C_{ei}(\tau)\} = \frac{1}{T_n} \int_{-\infty}^{\infty} B_i(\xi) \cdot B_i(\xi+\tau)$$
$$\cdot \left[\lim_{T\to\infty} \frac{1}{T} \int_{\xi-(T/2)}^{\xi+(T/2)} X(t) \cdot X(t+\tau) \cdot dt \right] d\xi.$$

The quantity in brackets is the true autocovariance function $C(\tau)$. Hence,

$$\text{ave } \{C_{ei}(\tau)\} = D_{ei}(\tau) \cdot C(\tau) \tag{B-4.10}$$

where

$$D_{ei}(\tau) = \frac{1}{T_n} \int_{-\infty}^{\infty} B_i(\xi) \cdot B_i(\xi+\tau) \cdot d\xi \tag{B-4.11}$$

is the *lag window equivalent* of the data window $B_i(t)$. Therefore,

$$\text{ave } \{P_{ei}(f)\} = Q_{ei}(f) * P(f) \tag{B-4.12}$$

where $Q_{ei}(f)$ is the *spectral window* corresponding to (i.e. the Fourier transform of) the lag window $D_{ei}(\tau)$.

If $J_i(f)$ is the *frequency window* corresponding to (i.e. the Fourier transform of) the data window $B_i(t)$, then

$$Q_{ei}(f) = \frac{1}{T_n} \mid J_i(f) \mid^2. \tag{B-4.13}$$

(It will be noted that (B-4.10) can be obtained more directly from (B-4.9), by taking the ensemble average of the right-hand member of (B-4.9). In this case, λ is superfluous and need not have been introduced at the start.)

These formulas for the "direct" method, where Fourier transformation (here modified by the data window) precedes multiplication (here squaring) can be used in several ways. Let us consider three possibilities:

(a) we may choose $B_i(t)$ non-zero over as long a range as possible, so that $D_{ei}(\tau)$ will be broad and $Q_{ei}(f)$ narrow,

(b) we may choose $B_i(t)$ non-zero over a much shorter range, making $D_{ei}(\tau)$ not so broad and $Q_{ei}(f)$ not so narrow,

(c) we may choose a number of such short windows, use each for $B_i(t)$ and average the corresponding values of $P_{ei}(f)$ so obtained into a general average.

If we follow choice (a), our result will behave similarly to those obtained from the indirect method when we try to make $Q_i(f)$ like $\delta(f)$. We shall estimate an average over a very short frequency interval, and our estimate will be excessively variable.

If we follow choice (b), we shall estimate an average over a wider frequency interval, but our estimate will remain just as variable.

If we follow choice (c) our estimate will refer to the same sort of smoothing as in (b), but we shall gain increased stability for the estimate. The behavior of the estimate will resemble that of a reasonable estimate by the indirect route.

Finally, there is another way in which we can apply the formulas for the direct route. We may use a long data window, calculate many values of $P_{ei}(f)$, and then average these results over moderately wide frequency intervals. Again our estimates will be estimates of considerably smoothed spectral densities; again our estimates will be moderately stable.

Of all these, the simplest description of an estimate which is moderately stable, and must, consequently, be an estimate of an at least moderately smoothed spectral density, is the indirect route. We shall stick to the indirect route for the present.

B.5 PARTICULAR PAIRS OF WINDOWS

In this section we will consider five pairs of windows. They are illustrated in Figs. 1, 14, and 15. We begin with the

Zeroth Pair

$$D_0(\tau) = 1, \qquad |\tau| < T_m,$$
$$= 0, \qquad |\tau| > T_m,$$

and

$$Q_0(f) = 2T_m \frac{\sin 2\pi f T_m}{2\pi f T_m} = 2T_m \text{ dif } 2f T_m.$$

Notice that $C_0(\tau) = D_0(\tau) \cdot C_{00}(\tau)$ coincides with $C_{00}(\tau)$ wherever $C_{00}(\tau)$

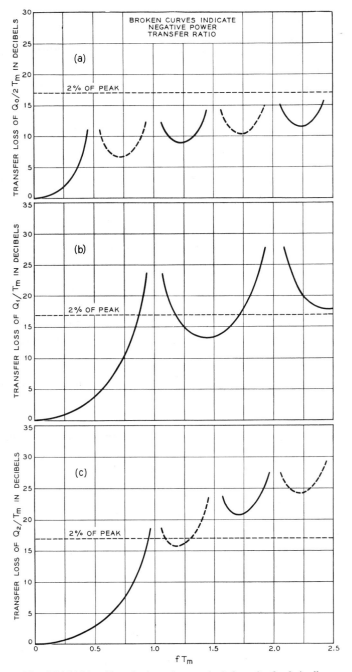

Fig. 15(a)(b)(c) — Transfer loss of spectral windows Q_0, Q_1, Q_2 in db.

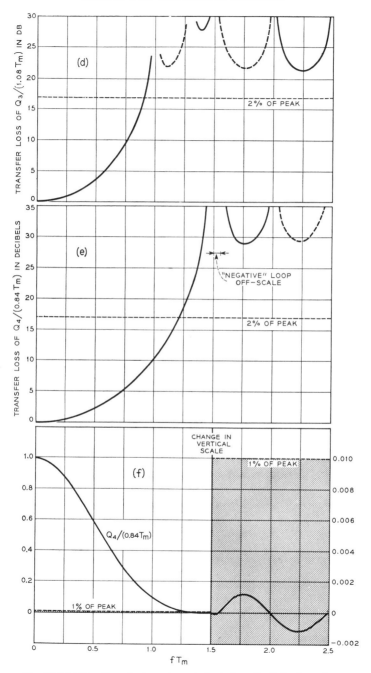

Fig. 15(d)(e)(f) — Transfer loss of spectral windows Q_3, Q_4 in db and power transfer ratio of spectral window Q_4.

is defined, and vanishes elsewhere. This is a "do-nothing" pair. However, $Q_0(f)$ is neither simple nor well-behaved, the first side lobe on each side of the main lobe being about $\frac{1}{5}$ the height of the main lobe (and negative).

The specifications of the other four pairs are:

First Pair (Bartlett[5])

$$D_1(\tau) = 1 - \frac{|\tau|}{T_m}, \qquad |\tau| < T_m,$$

$$= 0, \qquad |\tau| > T_m,$$

and

$$Q_1(f) = T_m \left(\frac{\sin \pi f T_m}{\pi f T_m} \right)^2.$$

Second Pair (sometimes called "hanning", after the Austrian meteorologist Julius von Hann)

$$D_2(\tau) = \frac{1}{2} \left(1 + \cos \frac{\pi \tau}{T_m} \right), \qquad |\tau| < T_m,$$

$$= 0, \qquad |\tau| > T_m,$$

and

$$Q_2(f) = \frac{1}{2} Q_0(f) + \frac{1}{4} \left[Q_0 \left(f + \frac{1}{2T_m} \right) + Q_0 \left(f - \frac{1}{2T_m} \right) \right].$$

Third Pair (sometimes called "hamming", after R. W. Hamming[26])

$$D_3(\tau) = 0.54 + 0.46 \cos \frac{\pi \tau}{T_m}, \qquad |\tau| < T_m,$$

$$= 0, \qquad |\tau| > T_m,$$

and

$$Q_3(f) = 0.54 Q_0(f) + 0.23 \left[Q_0 \left(f + \frac{1}{2T_m} \right) + Q_0 \left(f - \frac{1}{2T_m} \right) \right].$$

Fourth Pair (RBB's not **very** serious proposal)

$$D_4(\tau) = 0.42 + 0.50 \cos \frac{\pi \tau}{T_m} + 0.08 \cos \frac{2\pi \tau}{T_m}, \qquad |\tau| < T_m,$$

$$= 0, \qquad |\tau| > T_m$$

and

$$Q_4(f) = 0.42Q_0(f) + 0.25\left[Q_0\left(f + \frac{1}{2T_m}\right) + Q_0\left(f - \frac{1}{2T_m}\right)\right]$$
$$+ 0.04\left[Q_0\left(f + \frac{1}{T_m}\right) + Q_0\left(f - \frac{1}{T_m}\right)\right].$$

These specifications are all special cases of

$$D_1(\tau) = a_{i0} + 2\sum_{j=1}^{\infty} a_{ij}\cos\frac{j\pi\tau}{T_m}, \qquad |\tau| < T_m,$$
$$= 0, \qquad\qquad\qquad |\tau| > T_m,$$

with

$$a_{i0} + 2\sum_{j=1}^{\infty} a_{ij} = 1,$$

whence,

$$Q_i(f) = a_{i0}Q_0(f) + \sum_{j=1}^{\infty} a_{ij}\left[Q_0\left(f + \frac{j}{2T_m}\right) + Q_0\left(f - \frac{j}{2T_m}\right)\right].$$

The coefficients in $D_3(\tau)$ may be regarded as convenient approximations to

$$a_{30} = \frac{25}{46}, \qquad a_{31} = \frac{21}{92},$$

which would have produced a zero of $Q_3(f)$ at $|f| = 1.25/T_m$, with other zeros occurring at all integral multiples of $0.5/T_m$ except at 0 and $\pm 0.5/T_m$. (The zero which could have been produced at $|f| = 1.25/T_m$ actually occurs at approximately $|f| = 1.3/T_m$.) The coefficients in $D_3(\tau)$ were actually selected to minimize the height of the highest side lobe.

The coefficients in $D_4(\tau)$ are convenient approximations to

$$a_{40} = \frac{3969}{9304}, \qquad a_{41} = \frac{1155}{4652}, \qquad a_{42} = \frac{715}{18608},$$

which would have produced zeros of $Q_4(f)$ at $|f| = 1.75/T_m$ and $|f| = 2.25/T_m$, with other zeros occurring at all integral multiples of $0.5/T_m$ except at 0, $\pm 0.5/T_m$, and $\pm 1/T_m$.

In view of the fact that

$$D_i(\tau) \cdot D_0(\tau) = D_i(\tau),$$

we have

$$Q_i(f) * Q_0(f) = Q_i(f),$$

relations which may be of interest, as may the fact that using $D_i(\tau) \cdot D_j(\tau)$ for the lag window corresponds in general, to a spectral window $Q_i(f) * Q_j(f)$. The writers have spent considerable time and effort inquiring into other possible window pairs. One of the most promising approaches was that of Čebyšěv or Chebyshëv polynomials (see Dolph[27]) to obtain side lobes of equal height. Their present conviction is that: (i) special windows *cannot* eliminate the need for prewhitening and rejection filtration and (ii) good prewhitening and rejection filtration *can* eliminate the need for special windows. Accordingly they do not recommend expending extensive effort on special windows.

Readers familiar with physical optics will recognize the close relation between the considerations of this section and diffraction by slits of uniform ($i = 0$) or varying ($i = 1, 2, 3, 4$) width. The literature on apodization (Jaquinot,[28] Boughon, Dossier, Jaquinot[29]) is relevant.

B.6 COVARIABILITY OF POWER DENSITY ESTIMATES — BASIC RESULT

To derive a formula for the covariance of two power density estimates $P_i(f_1)$, $P_j(f_2)$ obtained from the same record, we will first derive a formula for the covariance of $M(t_1, \tau_1)$, $M(t_2, \tau_2)$ where

$$M(t, \tau) = X\left(t - \frac{\tau}{2}\right) \cdot X\left(t + \frac{\tau}{2}\right).$$

For this we will use a formula for

$$\text{cov}\{wx, yz\} \equiv \text{ave}\{wxyz\} - \text{ave}\{wx\}\,\text{ave}\{yz\}$$

dating back to Isserlis,[30] and used by Hotelling[31] in a similar connection. If w, x, y, z are joint Gaussian variates with zero averages, then

$$\text{cov}\{wx, yz\} = \text{ave}\{wy\} \cdot \text{ave}\{xz\} + \text{ave}\{wz\} \cdot \text{ave}\{xy\}.$$

(This result is easily derived from the "characteristic function".) If we take

$$w = X\left(t_1 - \frac{\tau_1}{2}\right), \qquad x = X\left(t_1 + \frac{\tau_1}{2}\right),$$

$$y = X\left(t_2 - \frac{\tau_2}{2}\right), \qquad z = X\left(t_2 + \frac{\tau_2}{2}\right),$$

and make use of

$$\text{ave}\left\{X\left(t-\frac{\tau}{2}\right)\cdot X\left(t+\frac{\tau}{2}\right)\right\} = C(\tau) = \int_{-\infty}^{\infty} P(f)\cdot e^{i\omega\tau}\, df$$

we get

$$\text{cov}\,\{M(t_1,\tau_1), M(t_2,\tau_2)\} =$$

$$\iint\limits_{-\infty}^{\infty} [e^{i(\omega_1-\omega_2)(\tau_1-\tau_2)/2} + e^{i(\omega_1-\omega_2)(\tau_1+\tau_2)/2}] \tag{B-6.1}$$

$$\cdot e^{-i(\omega_1+\omega_2)(t_1-t_2)}\cdot P(f_1)\cdot P(f_2)\cdot df_1\cdot df_2.$$

It may be noted that

$$\text{var}\,\{[X(t)]^2\} = \text{cov}\,\{M(t,0), M(t,0)\} = 2\left[\int_{-\infty}^{\infty} P(f)\cdot df\right]^2,$$

while

$$\text{ave}\,\{[X(t)]^2\} = \int_{-\infty}^{\infty} P(f)\cdot df.$$

Hence,

$$\frac{2\cdot[\text{ave}\,\{[X(t)]^2\}]^2}{\text{var}\,\{[X(t)]^2\}} = 1,$$

in accordance with the fact that $[X(t)]^2$ is a constant multiple of a chi-square variate with one degree of freedom.

Substituting $f_1 = f' + f, f_2 = f' - f$, and noting that $df_1\cdot df_2 = 2df\cdot df'$, we get

$$\text{cov}\,\{M(t_1,\tau_1), M(t_2,\tau_2)\} =$$

$$\frac{1}{2}\int_{-\infty}^{\infty} [e^{i\omega(\tau_1-\tau_2)} + e^{i\omega(\tau_1+\tau_2)}]\cdot \Phi(f, t_1 - t_2)\cdot df, \tag{B-6.2}$$

where

$$\Phi(f,\lambda) = 4\int_{-\infty}^{\infty} P(f'+f)\cdot P(f'-f)\cdot e^{-i2\omega'\lambda}\, df'. \tag{B-6.3}$$

Since $\Phi(f,\lambda)$ is an even function of f we may replace $e^{i\omega(\tau_1\pm\tau_2)}$ by $\cos\omega(\tau_1\pm\tau_2)$, and since $P(f'+f)\cdot P(f'-f) = P(f+f')\cdot P(f-f')$ is an even function of f' we may replace $e^{-i2\omega'\lambda}$ by $\cos 2\omega'\lambda$. Hence, we have

$$\text{cov}\,\{M(t_1, \tau_1), M(t_2, \tau_2)\} =$$

$$\int_{-\infty}^{\infty} \cos \omega\tau_1 \cdot \cos \omega\tau_2 \cdot \Phi(f, t_1 - t_2) \cdot df, \quad \text{(B-6.4)}$$

where

$$\Phi(f, \lambda) = 4 \int_{-\infty}^{\infty} P(f + f') \cdot P(f - f') \cdot \cos 2\omega'\lambda \cdot df'. \quad \text{(B-6.5)}$$

The next step is to determine the covariance between $C_0(\tau_1)$ and $C_0(\tau_2)$. By definition, and for hypothetical computation,

$$C_0(\tau) = \frac{1}{T_n - |\tau|} \int_{-(T_n - |\tau|)/2}^{(T_n - |\tau|)/2} M(t, \tau) \cdot dt, \quad \text{(B-6.6)}$$

but this is inconvenient for present purposes on account of the dependence of both the limits of integration and the divisor upon τ. A more convenient form would be

$$C_0'(\tau) = \frac{1}{T_n'} \int_{-T_n'/2}^{T_n'/2} M(t, \tau) \cdot dt, \quad \text{(B-6.7)}$$

where $T_n' \leq T_n - T_m$. This form could actually be used for computation, but it would not make the maximum possible use of the data. The range of integration in $C_0'(\tau)$ is less than it is in $C_0(\tau)$ for any τ except possibly $|\tau| = T_m$. A good approximation to the use of $C_0(\tau)$ for computation is to regard this as (approximately) equivalent to the use of a hypothetical, modified $C_0'(\tau)$ with $T_n - T_m < T_n' < T_n$. We will take

$$T_n' = T_n - \alpha_i T_m, \quad \text{(B-6.8)}$$

where α_i depends upon the lag window to be used. Since, for each value of τ, the range of integration in formula (B-6.6) suffers a loss of τ out of T_n, and since the seriousness of this loss depends on the value of $D_i(\tau)$, it seems to be a reasonable approximation to take

$$\alpha_i = \frac{1}{T_m} \frac{\int_0^{T_m} \tau \cdot D_i(\tau) \cdot d\tau}{\int_0^{T_m} D_i(\tau) \cdot d\tau}, \quad \text{(B-6.9)}$$

which yields, for the first four lag windows described in Section B.5, $\alpha_0 = 0.50$, $\alpha_1 = 0.33$, $\alpha_2 = 0.30$, $\alpha_3 = 0.33$.

Using the approximation described in the preceding paragraph, we find (omitting the prime in C_0')

$$\text{cov}\,\{C_0(\tau_1),\, C_0(\tau_2)\} = \int_{-\infty}^{\infty} \cos\omega\tau_1\cdot\cos\omega\tau_2\cdot\Gamma(f)\,df \quad \text{(B-6.10)}$$

where

$$\Gamma(f) = \frac{1}{(T_n')^2} \iint\limits_{-T_n'/2}^{T_n'/2} \Phi(f,\, t_1 - t_2)\cdot dt_1\cdot dt_2$$

$$= 4\int_{-\infty}^{\infty} P(f + f')\cdot P(f - f')\cdot\left(\frac{\sin\omega' T_n'}{\omega' T_n'}\right)^2\cdot df'. \quad \text{(B-6.11)}$$

The final step is to determine the covariance of $P_i(f_1)$ with $P_j(f_2)$, recalling that, for example,

$$P_i(f) = \int_{-\infty}^{\infty} C_i(\tau)\cdot\cos\omega\tau\cdot d\tau,$$

where $C_i(\tau) = D_i(\tau)\cdot C_0(\tau)$. We get

$$\text{cov}\,\{P_i(f_1),\, P_j(f_2)\} = \frac{1}{4}\int_{-\infty}^{\infty} H_i(f;\, f_1)\cdot H_j(f;\, f_2)\cdot\Gamma(f)\cdot df, \quad \text{(B-6.12)}$$

where

$$H_i(f;\, f_1) = 2\int_{-\infty}^{\infty} D_i(\tau_1)\cdot\cos\omega\tau_1\cdot\cos\omega_1\tau_1\cdot d\tau_1$$

$$= Q_i(f + f_1) + Q_i(f - f_1) \quad \text{(B-6.13)}$$

with a similar formula for $H_j(f;\, f_2)$. In particular, of course,

$$\text{var}\,\{P_i(f_1)\} = \frac{1}{4}\int_{-\infty}^{\infty} [H_i(f;\, f_1)]^2\cdot\Gamma(f)\cdot df. \quad \text{(B-6.14)}$$

The *power-variance spectrum* is given by (B-6.11), and involves the true power spectrum in an essential way, as we would expect. Together with (B-6.12) and (B-6.14), this is the basic result. It is exact for estimates based on $C_0'(\tau)$, approximate for estimates based on $C_0(\tau)$.

B.7   COVARIABILITY OF ESTIMATES — VARIOUS APPROXIMATIONS

We now need to obtain a variety of approximate forms of the basic result, suitable for use under different conditions. We shall, in turn; (i) assume that the typical frequency scale of $H_i(f;\, f_1)$ and $H_j(f;\, f_2)$ is much larger than $1/T'_n$, (ii) assume that $P(f)$ varies slowly enough for the *quadratic* term in its Taylor series to be neglected at distances up to, say, $3/T'_n$, (iii) assume both (i) and an assumption similar to (ii) about $P_{i1}(f)$ and $P_{j2}(f)$, and, finally, (iv) assume only that $P(f)$ is concentrated in a sharp peak, narrow in comparison with $1/T'_n$.

Combining formulas (B-6.11) and (B-6.12) just obtained, we find

$$\text{cov } \{P_i(f_1), P_j(f_2)\} =$$

$$\iint\limits_{-\infty}^{\infty} H_i(f;\, f_1) \cdot H_j(f;\, f_2) \cdot P(f + f') \cdot P(f - f') \qquad \text{(B-7.1)}$$

$$\left(\frac{\sin \omega' T'_n}{\omega' T'_n} \right)^2 \cdot df \cdot df'.$$

The last factor in the integrand becomes rapidly negligible for $|f'|$ greater than, say, $1/T'_n$. On the other hand, the typical frequency scale of $H_i(f; f_1)$ and $H_j(f; f_2)$ is $1/T_m$ which is usually much larger than $1/T'_n$. In most cases, then, we will find that the approximation

$$H_i(f; f_1) \cdot H_j(f; f_2) \approx H_i(f + f'; f_1) \cdot H_j(f - f'; f_2)$$

is a good approximation for the values of f' making contributions of any importance to the integral. Under this approximation, and a change of variables of integration to

$$f'_1 = f + f', \qquad f'_2 = f - f',$$

we have, approximately,

$$\text{cov } \{P_i(f_1), P_j(f_2)\} \approx$$

$$\frac{1}{2} \int_{-\infty}^{\infty} H_i(f'_1;\, f_1) \cdot P(f'_1) \cdot H_j(f'_2;\, f_2) \cdot P(f'_2)$$

$$\cdot \left[\frac{\sin \pi(f'_1 - f'_2) T'_n}{\pi(f'_1 - f'_2) T'_n} \right]^2 \cdot df'_1 \cdot df'_2.$$

If we let

$$P_{i1}(f) = H_i(f; f_1) \cdot P(f), \qquad \text{(B-7.2)}$$

and

$$P_{j2}(f) = H_j(f; f_2) \cdot P(f), \qquad \text{(B-7.3)}$$

then, approximately,

$$\text{cov}\{P_i(f_1),\, P_j(f_2)\}$$

$$\approx 2 \iint\limits_{0}^{\infty} P_{i1}(f_1')\cdot P_{j2}(f_2')\cdot\left[\frac{\sin \pi(f_1' - f_2')T_n'}{\pi(f_1' - f_2')T_n'}\right]^2\cdot df_1'\cdot df_2', \qquad (B\text{-}7.4)$$

which is our approximation on assumption (i).

In terms of the same quantities we have, from (B-4.5),

$$\text{ave}\{P_i(f_1)\} = \int_0^\infty P_{i1}(f_1')\cdot df_1', \qquad (B\text{-}7.5)$$

and

$$\text{ave}\{P_j(f_2)\} = \int_0^\infty P_{j2}(f_2')\cdot df_2'. \qquad (B\text{-}7.6)$$

The following heuristic interpretation of the last three formulas is useful. Frequency f_1' is involved in $P_i(f_1)$ with a net weight of $P_{i1}(f_1')$, while frequency f_2' is involved in $P_j(f_2)$ with a net weight of $P_{j2}(f_2')$. These are net weights, and might represent partial cancellation. The covariance between $P_i(f_1)$ and $P_j(f_2)$ might involve additional sources of variability. To the extent that the approximation to $H_i(f; f_1)\cdot H_j(f; f_2)$ is valid, there are no additional sources of variability — the covariance involves the same net weights. The fact that we have only a record of equivalent length T_n' is represented by a tendency of the frequency f_1' to become entwined with the frequency f_2', measured by the factor

$$\left[\frac{\sin \pi(f_1' - f_2')T_n'}{\pi(f_1' - f_2')T_n'}\right]^2.$$

The fact that there are no additional sources of variability as we pass from first to second moments (to the accuracy of the approximation) is good evidence that we are using the data in a relatively efficient way.

If we begin anew from (B-7.1) by expanding $P(f \pm f')$ in Taylor series around $f' = 0$, we find

$$P(f + f')\cdot P(f - f') = [P(f)]^2 + (f')^2\{P(f)\cdot P''(f) - [P'(f)]^2\} + \cdots\cdots$$

symmetry forcing the odd-order terms, including the first, to vanish. Since the trigonometric factor is very small for $|f'| > 2$ or 3 times $1/T_n'$, we may often replace $P(f + f')\cdot P(f - f')$ by $[P(f)]^2$ to a very good approximation. This yields, after integrating out f',

$$\Gamma(f) \approx \frac{2}{T_n'}\, [P(f)]^2,$$

whence,

$$\operatorname{cov}\{P_i(f_1), P_j(f_2)\} \approx \frac{1}{T'_n} \int_0^\infty H_i(f; f_1) \cdot H_j(f; f_2)[P(f)]^2 \cdot df$$

$$\approx \frac{1}{T'_n} \int_0^\infty P_{i1}(f) P_{j2}(f) \, df,$$

our useful approximation on assumption (ii).

In carrying out this approximation we only needed smoothness of $P(f)$ for f's which make a non-negligible contribution to the final result. If $H_i(f; f_1)P(f)$ and $H_j(f; f_2)P(f)$ are both smooth, as under assumption (iii), then $P(f)$ must be smooth except where both $H_i(f; f_1)$ and $H_j(f; f_2)$ are small, and the contribution from such regions can be neglected. Thus our useful approximation for the *covariance* under assumptions (iii) is the same as that under assumption (ii). (The approximation for $\Gamma(f)$ need not be so accurate.)

Finally, if $P(f)$ consists only of a very sharp peak (width $\ll 1/T'_n$) at $f = f_0$ (and, of course, at $f = -f_0$), with area A, then

$$P(f) \approx A \cdot [\delta(f + f_0) + \delta(f - f_0)],$$

whence,

$$\Gamma(f) \approx 2A^2 \cdot \left[\delta(f + f_0) + \delta(f - f_0) + 2\left(\frac{\sin \omega_0 T'_n}{\omega_0 T'_n} \right)^2 \cdot \delta(f) \right].$$

Hence,

$$\operatorname{cov}\{P_i(f_1), P_j(f_2)\} \approx A^2 \left\{ H_i(f_0 \; ; f_1) \cdot H_j(f_0 \; ; f_2) \right.$$

$$\left. + H_i(0; f_1) \cdot H_j(0; f_2) \cdot \left(\frac{\sin \omega_0 T'_n}{\omega_0 T'_n} \right)^2 \right\},$$

which is our useful approximation under assumption (iv). In case $f_0 \gg 1/T'_n$, the trigonometric factor may be neglected, and we may write

$$\operatorname{cov}\{P_i(f_1), P_j(f_2)\} \approx \left(\int_0^\infty P_{i1}(f) \, df \right) \cdot \left(\int_0^\infty P_{j2}(f) \, df \right).$$

B.8 EQUIVALENT WIDTHS

Under assumption (ii) of Section B.7, i.e. $P(f)$ slowly varying, we have, for the dimensionless variability of $P_i(f_1)$,

$$\frac{\operatorname{var}\{P_i(f_1)\}}{[\operatorname{ave}\{P_i(f_1)\}]^2} = \frac{1}{T'_n W_e}, \tag{B-8.1}$$

where

$$W_e = \frac{\left[\int_0^\infty P_{i1}(f) \cdot df\right]^2}{\int_0^\infty [P_{i1}(f)]^2 \cdot df} \qquad \text{(B-8.2)}$$

is the *equivalent width* of

$$P_{i1}(f) = H_i(f; f_1) \cdot P(f).$$

We will now determine the equivalent width of $P_{i1}(f)$ under the assumption that, for each value of f_1, $P_{i1}(f)$ is essentially a constant times $H_i(f; f_1)$. The value of the constant may depend upon the value of f_1, but it will not have any effect on the value of W_e. It is convenient to express (B-8.2) in terms of the normalized frequency

$$\varphi = 2\pi f T_m, \qquad (\varphi_1 = 2\pi f_1 T_m),$$

so that

$$W_e = \frac{1}{4\pi T_m} \frac{\left[\int_{-\infty}^\infty \tilde{P}_{i1}(\varphi) \cdot d\varphi\right]^2}{\int_{-\infty}^\infty [\tilde{P}_{i1}(\varphi)]^2 \cdot d\varphi}, \qquad \text{(B-8.3)}$$

where

$$\left.\begin{aligned}
\tilde{P}_{i1}(\varphi) &= \tilde{Q}_i(\varphi + \varphi_1) + \tilde{Q}_i(\varphi - \varphi_1), \\
\tilde{Q}_i(\varphi) &= (1 - a_i)\tilde{Q}_0(\varphi) + \frac{a_i}{2}[\tilde{Q}_0(\varphi + \pi) + \tilde{Q}_0(\varphi - \pi)], \\
\tilde{Q}_0(\varphi) &= \frac{\sin \varphi}{\varphi}, \\
a_0 &= 0, \qquad a_2 = 0.50, \qquad a_3 = 0.46.
\end{aligned}\right\} \qquad \text{(B-8.4)}$$

Since

$$\int_{-\infty}^\infty \frac{\sin \varphi}{\varphi} \, d\varphi = \pi,$$

we get

$$\int_{-\infty}^\infty \tilde{P}_{i1}(\varphi) \cdot d\varphi = 2\pi,$$

and, since

$$\int_{-\infty}^\infty \frac{\sin (\varphi + \alpha)}{\varphi + \alpha} \frac{\sin (\varphi + \beta)}{\varphi + \beta} \, d\varphi = \frac{\sin (\alpha - \beta)}{\alpha - \beta} \pi,$$

we get

$$\int_{-\infty}^{\infty} [\hat{P}_{i1}(\varphi)]^2 \, d\varphi = 2\pi B,$$

where

$$B = \left(1 - 2a_i + \frac{3}{2} a_i^2\right)\left(1 + \frac{\sin 2\varphi_1}{2\varphi_1}\right)$$

$$+ a_i(1 - a_i)\left[\frac{\sin (2\varphi_1 - \pi)}{2\varphi_1 - \pi} + \frac{\sin (2\varphi_1 + \pi)}{2\varphi_1 + \pi}\right] \qquad \text{(B-8.5)}$$

$$+ \frac{a_i^2}{4}\left[\frac{\sin 2(\varphi_1 - \pi)}{2(\varphi_1 - \pi)} + \frac{\sin 2(\varphi_1 + \pi)}{2(\varphi_1 + \pi)}\right].$$

Hence, for these windows,

$$W_e = \frac{1}{2T_m B}. \qquad \text{(B-8.6)}$$

It will be noted that

$$B = 2\left(1 - 2a_i + \frac{3}{2} a_i^2\right), \qquad \text{for} \quad f_1 = 0,$$

$$= 1 - a_i + \frac{1}{2} a_i^2, \qquad \text{for} \quad f_1 T_m = \frac{1}{4},$$

$$= 1 - 2a_i + \frac{7}{4} a_i^2, \qquad \text{for} \quad f_1 T_m = \frac{1}{2},$$

$$= 1 - 2a_i + \frac{3}{2} a_i^2, \qquad \text{for} \quad f_1 T_m = \frac{r}{4},$$

$$(r = 3, 4, 5, \cdots).$$

To a close approximation,

$$B = 1 - 2a_i + \frac{3}{2} a_i^2, \qquad \text{for} \quad f_1 T_m \geqq 1,$$

whence, for

$i = 0$, the main lobe is $1/T_m$ wide, and the equivalent width of $P_{01}(f)$ is $1/(2T_m)$,

$i = 2$, the main lobe is $2/T_m$ wide, and the equivalent width of $P_{21}(f)$ is $4/(3T_m)$,

$i = 3$, the main lobe is $2/T_m$ wide, and the equivalent width of $P_{31}(f)$ is $1.258/T_m$.

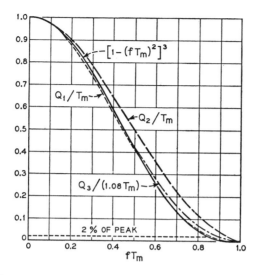

Fig. 16 — Approximation to standard spectral windows.

For $f_1 < 1/T_m$ these equivalent widths are to be reduced. For $f_1 = 0$, they are to be halved.

Assume, next, that

$$Q_i(f) = A \, [1 - (T_m f)^2]^3, \qquad |f| \leqq \frac{1}{T_m},$$

$$= 0, \qquad \text{otherwise,} \tag{B-8.7}$$

which approximates $Q_i(f)$ quite well for $i = 1, 2,$ and 3, as is shown in Fig. 16. Let us further assume that $f_1 \geqq 1/T_m$ and that, at least for $|f - f_1| \leqq 1/T_m$,

$$P(f) = P(f_1)[1 + \beta T_m(f - f_1)]. \tag{B-8.8}$$

Then we can evaluate the equivalent width of

$$P_{i1}(f) = H_i(f; f_1)P(f)$$

by simple integrations, finding

$$W_e = \frac{429}{350 T_m \left(1 + \dfrac{\beta^2}{15}\right)}.$$

Since β^2 cannot exceed 1, we have

$$\frac{1.15}{T_m} < W_e < \frac{1.23}{T_m}, \tag{B-8.9}$$

showing how little effect a linear slope of $P(f)$ across the main lobe of $H_i(f; f_1)$ has on W_e, even when β is large enough for $P(f)$ to vanish at one edge.

If $P(f)$ has exactly the same form as the $H_i(f; f_1)$ corresponding to (B-8.7), then

$$W_e = \frac{0.94}{T_m},$$ (B-8.10)

a rather smaller value.

If, on the other hand, we take

$$Q_i(f) = A[1 - (T_m f)^2]^\ell, \quad |f| \geqq \frac{1}{T_m},$$ (B-8.11)

$$= 0, \quad \text{otherwise,}$$

then for $\ell = 2$, and $P(f)$ given by (B-8.8),

$$W_e = \frac{7}{5T_m\left(1 + \dfrac{\beta^2}{11}\right)},$$

so that

$$\frac{1.28}{T_m} < W_e \leqq \frac{1.40}{T_m},$$ (B-8.12)

while for $\ell = 1$,

$$W_e = \frac{5}{3T_m\left(1 + \dfrac{\beta^2}{7}\right)},$$

so that

$$\frac{1.46}{T_m} < W_e < \frac{1.67}{T_m}.$$ (B 8.13)

In general, we can use

$$W_e \approx \frac{1}{T_m}$$ (B-8.14)

as a conservative approximation which provides a factor of safety often near 1.15 or 1.20.

All this was for $f_1 \geqq 1/T_m$. As f_1 is reduced below $1/T_m$ there is overlapping between $Q_i(f - f_1)$ and $Q_i(f + f_1)$ in $H_i(f; f_1)$. As a consequence, the equivalent width decreases in a way which is not worth examining in great detail. (At $f_1 = 1/(2T_m)$ the equivalent width has decreased about 15 per cent, and at $f_1 = 0$ it has fallen to just half its usual value.)

B.9 EQUIVALENT DEGREES OF FREEDOM

Let us assume that $P(f)$ is flat — uniformly equal to p_0. Then the power variance spectrum $\Gamma(f)$ will also be flat, but with a value of $2p_0^2/T_n'$. Let us consider $H_i(f; f_1)$ to be the ideal bandpass power transfer function

$$H_i(f; f_1) = A, \qquad f_1 - \frac{W}{2} < |f| < f_1 + \frac{W}{2},$$

$$= 0, \qquad \text{otherwise},$$

where $f_1 > W/2$, although such a transfer function is not even approximately realizable under the requirement that $D_i(\tau) \equiv 0$ for $|\tau| > T_m$. Then

$$\text{ave } \{P_i(f_1)\} = AWp_0,$$

and

$$\text{var } \{P_i(f_1)\} = \frac{A^2 W p_0^2}{T_n'}.$$

If we equate these moments with the corresponding moments of a multiple of a chi-square variate with k degrees of freedom, we get

$$k = \frac{2[\text{ave } \{P_i(f_1)\}]^2}{\text{var } \{P_i(f_1)\}} = 2WT_n'.$$

We get the same result if $H_i(f; f_1)$ is assumed to be the ideal lowpass power transfer function

$$H_i(f; f_1) = A, \qquad |f| < W,$$

$$= 0, \qquad \text{otherwise}.$$

In either case we get only one degree of freedom when $W = 1/(2T_n')$.

This suggests that the frequency range $f > 0$ be divided into *elementary bands* of width

$$\Delta f = \frac{1}{2T_n'} \tag{B-9.1}$$

with *one degree of freedom in each*. (In the presence of very sharp peaks in the original spectrum it would be somewhat more accurate to divide the frequency range $-\infty < f < \infty$ into bands of width $1/T_n'$ with one degree of freedom in each, and with one band centered at $f = 0$.) It follows from the results of the preceding section that, if $P(f)$ is reasonably smooth, we may regard the stability of the power density estimate $P_i(f_1)$, for $i = 1, 2,$ or 3, as approximately equal to that of a chi-square variate with k degrees of freedom, where

$$k = \frac{2T'_n}{T_m} = 2\left(\frac{T_n}{T_m} - \frac{1}{3}\right) \qquad \text{(B-9.2)}$$

for one piece (or record).

B.10 FILTERING AND ANALOG COMPUTATION

Power spectrum analysis from continuous data is frequently done by filtering techniques. In this section we will examine some of these techniques. In particular, we will try to express their results in such a form that we are led to express their reliability in equivalent numbers of degrees of freedom.

A common technique is to apply the signal to a narrow-band filter, allow some time for initial transients to become negligible, and then measure the average output power over the remaining time of the record. This corresponds to Mode I as described in Section 10 of Part I. In this technique it is clear that the estimate $P_Y(f_1)$ of the power density $P(f_1)$ in the power spectrum $P(f)$ of the original random process at the filter input, is, substantially,

$$P_Y(f_1) = \int_{-\infty}^{\infty} P_{\text{out}}(f; f_1) \cdot df$$

where

$$P_{\text{out}}(f; f_1) = |Y(f; f_1)|^2 \cdot P(f)$$

is the power spectrum of the modified random process at the filter output, and $Y(f; f_1)$ is the transfer function of the filter with a narrow passband around the frequency f_1. The record length is shorter for the modified process than for the original process, by the time allowed for the initial transients to become negligible (at least the reciprocal of the bandwidth in cycles). The effective record length for the modified process determines the width of the elementary bands, and the equivalent number of degrees of freedom is the number of elementary bands in the equivalent width of $P_{\text{out}}(f; f_1)$, or the equivalent width of $|Y(f; f_1)|^2$ if $P(f)$ is reasonably constant in the filter passband.

In the technique described as Mode II in Section 10 we integrate *all* of the power output of the filter, and divide by the length T of the original record to obtain the estimate $P_Y(f_1)$. The result may clearly be expressed in the form

$$P_Y(f_1) = \frac{1}{T} \int_{-\infty}^{\infty} \{ W(t; f_1) * [B(t) \cdot X(t)] \}^2 \, dt ,$$

where $W(t; f_1)$ is the impulse response of the filter, and $B(t) = 0$ for $t < 0$ and $t > T$, but is otherwise arbitrary. To reduce this to a familiar form we first write it out in detail as

$$P_Y(f_1) = \frac{1}{T} \iiint\limits_{-\infty}^{\infty} W(\tau_1 ; f_1) \cdot W(\tau_2 ; f_1) \cdot B(t - \tau_1) \cdot B(t - \tau_2) \cdot X(t - \tau_1)$$

$$\cdot X(t - \tau_2) \cdot d\tau_1 \, d\tau_2 \, dt ,$$

so that

$$\text{ave } \{ P_Y(f_1) \} = \frac{1}{T} \iiint\limits_{-\infty}^{\infty} W(\tau_1 ; f_1) \cdot W(\tau_2 ; f_1) \cdot B(t - \tau_1) \cdot B(t - \tau_2)$$

$$\cdot C(\tau_1 - \tau_2) \cdot d\tau_1 \, d\tau_2 \, dt .$$

Now,

$$C(\tau_1 - \tau_2) = \int_{-\infty}^{\infty} P(f) \cdot e^{-i\omega(\tau_1 - \tau_2)} \, df ,$$

while, if $J(f)$ is the Fourier transform of $B(t)$,

$$\int_{-\infty}^{\infty} B(t - \tau_1) \cdot B(t - \tau_2) \cdot dt$$

$$= \iiint\limits_{-\infty}^{\infty} J(f') \cdot J(f'') \cdot e^{i\omega'(t-\tau_1) + i\omega''(t-\tau_2)} \, df' \, df'' \, dt$$

$$= \iint\limits_{-\infty}^{\infty} J(f') \cdot J(f'') \cdot \delta(f' + f'') \cdot e^{-i(\omega'\tau_1 + \omega''\tau_2)} \, df' \, df''$$

$$= \int_{-\infty}^{\infty} |J(f')|^2 \cdot e^{-i\omega'(\tau_1 - \tau_2)} \, df' .$$

Hence,

$$\text{ave } \{ P_Y(f_1) \} = \frac{1}{T} \iiiint\limits_{-\infty}^{\infty} |J(f')|^2 \cdot W(\tau_1; f_1) \cdot W(\tau_2; f_1) \cdot P(f)$$

$$\cdot e^{i(\omega - \omega')(\tau_1 - \tau_2)} \, d\tau_1 \, d\tau_2 \, df' \, df .$$

Further, since $Y(f; f_1)$ is the Fourier transform of $W(t; f_1)$,

$$\text{ave}\{P_Y(f_1)\} = \frac{1}{T} \iint\limits_{-\infty}^{\infty} |J(f')|^2 \cdot |Y(f - f'; f_1)|^2 \cdot P(f) \cdot df' \, df.$$

Finally, therefore,

$$\text{ave } \{P_Y(f_1)\} = \int_0^{\infty} H_Y(f; f_1) \cdot P(f) \cdot df, \tag{B-10.1}$$

where

$$H_Y(f; f_1) = \frac{2}{T} |J(f)|^2 * |Y(f; f_1)|^2. \tag{B-10.2}$$

Since (B-10.1) is in the same form as (B-4.5), we may now apply the results of Sections B.8 and B.9. In particular, if $P(f)$ is reasonably smooth, we may regard the stability of the estimate $P_Y(f_1)$ as approximately equal to that of a chi-square variate with k degrees of freedom, where

$$k = 2T \frac{\left[\int_0^{\infty} H_Y(f; f_1) \cdot df\right]^2}{\int_0^{\infty} [H_Y(f; f_1)]^2 \cdot df}. \tag{B-10.3}$$

From the fact that $H_Y(f; f_1)$ is the convolution of the power transfer function of the filter with $(2/T)|J(f)|^2$ it is clear that the passband of $H_Y(f; f_1)$ is at least as wide as that of the wider of the filter and $(2/T)|J(f)|^2$. Hence, the resolving power of the filter method of power spectrum analysis is limited by the length T of available data, just as is that of any other method. If the filter passband is made narrower than $1/T$, say, not only do we gain very little in resolving power, but the stability of our power density estimates is then, at best, approximately that of a chi-square variate with only one or two degrees of freedom. To obtain a reasonable degree of stability the filter passband should be several to many times $1/T$ wide. The resolving power then depends largely on this width. Under these circumstances it should not make much difference which of the four modes described in Section 10 is used.

In Mode III, the output power may clearly be expressed in the form

$$P_Y(f_1) = \frac{1}{T} \int_{-\infty}^{\infty} [\tilde{B}(t) \cdot \{W(t; f_1) * [B(t) \cdot X(t)]\}]^2 \, dt,$$

where $\tilde{B}(t) = 0$ for $t < 0$ and $t > T$, but is arbitrary otherwise. The reduction of this to a familiar form follows closely that of the preceding case. The final result is that

$$\text{ave } \{P_Y(f_1)\} = \int_0^\infty H_Y(f; f_1) \cdot P(f) \cdot df, \qquad \text{(B-10.4)}$$

where

$$H_Y(f; f_1) = \frac{2}{T} \int_{-\infty}^\infty \left| \int_{-\infty}^\infty J(f'') \cdot Y(f'' - f; f_1) \right.$$
$$\left. \cdot \tilde{J}(f' - f'') \cdot df'' \right|^2 \cdot df'. \qquad \text{(B-10.5)}$$

The analysis of Mode IV differs from that of Mode III only in the specifications of $\tilde{J}(f)$. Indeed, the results for Modes I and II are also special cases of the result for Mode III. Mode II corresponds to $\tilde{J}(f) = \delta(f)$, while Mode I corresponds to $J(f) = \delta(f)$ as well as $\tilde{J}(f) = \delta(f)$.

It was noted in Section 10 that the zero of the input noise may not be quite at ground potential in Fig. 5. The discrepancy may be considered to produce a spurious *line* or *delta* component at zero frequency in the spectrum of the input noise. It will have no effect on the average output power in Mode I if $f_1 \neq 0$, because the spectral window, which is simply $| Y(f; f_1) |^2$ is opaque at zero frequency. In Mode II, however, the spectral window specified by (B-10.2) may not be sufficiently opaque at zero frequency. The effect on the average output power depends upon the value of $H_Y(0; f_1)$. If the spectral window $H_Y(f; f_1)$ were ideal in the sense that it has no side lobes, there would be no effect on the average output power for values of f_1 at least half of the width of the spectral window. This indicates the desirability of using a graded data window $B(t)$ in order to reduce the side lobes in the spectral window $H_Y(f; f_1)$, that is, essentially the side lobes in $(2/T) |J(f)|^2$. Since the latter window is necessarily positive for all values of f, reduction of side lobes here is not quite as easy as it is in the indirect computation technique described in Section B.4, where selection can be exercised directly on the lag window, and the spectral window can be negative at some frequencies.

Another filter technique frequently used in power spectrum analysis is to apply the available record to the filter as a periodic function with a period equal to the length of the record. If a data window is used, so that the periodic function applied to the filter is of the form $B(t) \cdot X(t)$ in the interval $0 < t < T$, then the power spectrum of the input function is a *line spectrum* with power concentrated at integral multiples of $1/T$

cps. The total power in this spectrum, expressed in a form which displays the distribution in frequency, is

$$\text{input power} = \frac{1}{T} \sum_{q=-\infty}^{q=\infty} P_{\text{in}}\left(\frac{q}{T}\right),$$

where, essentially,

$$P_{\text{in}}(f) = \frac{1}{T} \left| \int_{-\infty}^{\infty} B(t) \cdot X(t) \cdot e^{-i\omega t} \cdot dt \right|^2.$$

Hence, the output power of the filter, is

$$P_Y(f_1) = \frac{1}{T} \sum_{q=-\infty}^{q=\infty} \left| Y\left(\frac{q}{T}; f_1\right) \right|^2 \cdot P_{\text{in}}\left(\frac{q}{T}\right).$$

Aside from the fact that we are now dealing with sums instead of integrals, this analysis parallels very closely our analysis of Mode II. It should be noted, however, that while our summands are even functions of q we may not run our sums from $q = 0$ to $q = \infty$, and then double the result, unless we take only half of the first ($q = 0$) term. Hence, with this technique the equivalent number of degrees of freedom should be taken as

$$k = \frac{\left[\sum_{q=-\infty}^{q=\infty} \left| Y\left(\frac{q}{T}; f_1\right) \right|^2 \right]^2}{\sum_{q=-\infty}^{q=\infty} \left| Y\left(\frac{q}{T}; f_1\right) \right|^4}$$

provided that this turns out to give $k > 3$ or 4, say. It is easily seen that this formula discourages the use of this technique with a filter whose passband is only a few times $1/T$ wide.

B.11 PREWHITENING

The techniques required here are standard in communication engineering and do not require specific treatment. As these techniques are ordinarily used, the original stationary random process is in effect continuously acting on the input end of the prewhitening filter, so that the output of the filter may be regarded as another stationary random process for purposes of spectral analysis. If these techniques are used where it is practical to obtain only a finite length of the original process to apply to the input of the filter, we must discard an initial portion of the output, corresponding to the time required for the filter transients to die out, as well as all of the output after the input has ceased. These considerations are taken up in greater detail in Sections B.15 and 16 in connection with the analysis of equi-spaced discrete data.

DETAILS FOR EQUI-SPACED ANALYSIS

B.12 ALIASING

Let us consider a stationary random process whose autocovariance function $\tilde{C}(\tau)$, and power spectrum $\tilde{P}(f)$ are known exactly. If we now take only the values of $\tilde{C}(\tau)$ at uniformly spaced values of τ, viz.

$$\tau = 0, \ \pm \Delta \tau, \ \pm 2\Delta \tau, \ \cdots$$

we can, in principle, calculate a corresponding (aliased) power spectrum $\tilde{P}_a(f)$, by using the formula

$$\tilde{P}_a(f) = \int_{-\infty}^{\infty} [\nabla(\tau; \Delta\tau) \cdot \tilde{C}(\tau)] \cdot e^{-i\omega\tau} \, d\tau,$$

where $\nabla(\tau; \Delta\tau)$ is an infinite Dirac comb as defined in Section A.2. Making use of the results of Sections A.2 and A.3 we have

$$\tilde{P}_a(f) = A\left(f; \frac{1}{\Delta\tau}\right) * \tilde{P}(f),$$

or, explicitly,

$$\tilde{P}_a(f) = \sum_{q=-\infty}^{q=\infty} \tilde{P}\left(f - \frac{q}{\Delta\tau}\right).$$

Thus, if it happens that $\tilde{P}(f)$ is zero for $|f| > 1/(2\,\Delta\tau)$, as illustrated in Fig. 17(a), there will be no overlapping of the individual terms in $\tilde{P}_a(f)$, and the result of the summation will be as illustrated in Fig. 17(b). In this case, we can restore the original spectrum by multiplying $\tilde{P}_a(f)$ by a rectangular function according to the formula

$$\tilde{P}(f) = \tilde{P}_A(f) = \tilde{S}(f) \cdot \tilde{P}_a(f),$$

where

$$\tilde{S}(f) = 1, \quad |f| < \frac{1}{2\Delta\tau},$$

$$= 0, \quad |f| > \frac{1}{2\Delta\tau},$$

for, in the absence of such overlapping, $\tilde{P}_a(f) = \tilde{P}(f)$ for $|f| < 1/(2\Delta\tau)$. Hence, we will recover the original autocovariance function $\tilde{C}(\tau)$ from that for the discrete series by a convolution according to the formula

$$\tilde{C}(\tau) = \tilde{G}(\tau) * [\nabla(\tau; \Delta\tau) \cdot \tilde{C}(\tau)]$$

$$= \Delta\tau \cdot \sum_{q=-\infty}^{q=\infty} \tilde{G}(\tau - q\Delta\tau) \cdot \tilde{C}(q\Delta\tau),$$

where

$$\tilde{G}(\tau) = \frac{\sin\dfrac{\pi\tau}{\Delta\tau}}{\pi\tau} = \frac{1}{\Delta\tau} \cdot \mathrm{dif}\,\frac{\tau}{\Delta\tau}.$$

Now let us consider a second stationary random process whose auto-covariance function $C(\tau)$ happens to be related to $\tilde{C}(\tau)$ by

$$C(\tau) = G(\tau) \cdot \tilde{C}(\tau),$$

where $G(\tau)$ is unity at $\tau = 0$, $\pm\Delta\tau$, $\pm 2\Delta\tau$, etc. Since

$$\nabla(\tau; \Delta\tau) \cdot C(\tau) = \nabla(\tau; \Delta\tau) \cdot \tilde{C}(\tau),$$

it is clear that $P_a(f)$ will be quantitatively, and in the utmost detail,

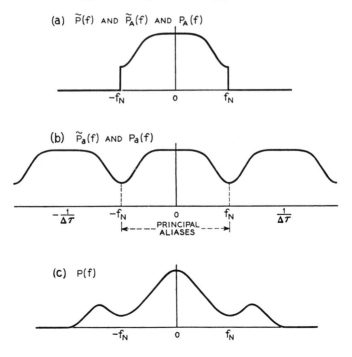

(a) $\tilde{P}(f)$ AND $\tilde{P}_A(f)$ AND $P_A(f)$

(b) $\tilde{P}_a(f)$ AND $P_a(f)$

(c) $P(f)$

Fig. 17 — Effect of sampling on power spectra.

identical with $\tilde{P}_a(f)$. Hence, the *principal part* $\tilde{P}_A(f)$ of the aliased spectrum $\tilde{P}_a(f)$, that is, the part in the range $|f| < 1/(2\Delta\tau)$, will not in general be the power spectrum of the second process. In fact, if $S(f)$ is the Fourier transform of $G(\tau)$, the power spectrum of the second process is related to the power spectrum of the first process by

$$P(f) = S(f) * \tilde{P}(f).$$

In general, this spectrum covers an infinite range of frequencies, so that the aliased spectrum

$$P_a(f) = A\left(f; \frac{1}{\Delta\tau}\right) * P(f) = \sum_{q=-\infty}^{\infty} P\left(f - \frac{q}{\Delta\tau}\right)$$

will involve overlapping of the individual terms. This overlapping will account for the quantitative identity of $P_a(f)$ with $\tilde{P}_a(f)$, and the failure of its principal part $P_A(f)$ to represent $P(f)$ even in the range $|f| < 1/(2\,\Delta\tau)$. Fig. 17(c) illustrates such a $P(f)$.

It may well have already occurred to readers familiar with amplitude modulation that using only uniformly spaced values of $C(\tau)$, viz., $C(r\Delta\tau)$ where $r = 0, \pm1, \pm2, \cdots$, has the same effect on the power spectrum as the simultaneous amplitude modulation of carrier waves with frequencies $q/\Delta\tau$ where $q = 0, \pm1, \pm2, \cdots$. If the two-sided power spectrum $P(f)$ corresponding to $C(\tau)$ is visualized as side-bands on a zero-frequency carrier, then the aliased spectrum $P_a(f)$ corresponding to $C(r\Delta\tau)$ will be naturally visualized as the same sidebands on carrier frequencies $q/\Delta\tau$, where $q = 0, \pm1, \pm2, \cdots$. If each sideband of $P(f)$ does not extend beyond the frequency $1/(2\Delta\tau)$, then there will be no overlapping of side-bands in $P_a(f)$, and the principal part $P_A(f)$ of $P_a(f)$ will be identical to $P(f)$. Contrariwise, if each sideband of $P(f)$ extends beyond the frequency $1/(2\Delta\tau)$, then there will be overlapping of side-bands in $P_a(f)$, and the principal part $P_A(f)$ of $P_a(f)$ will not be identical to $P(f)$. In any case, however, it is important to note that

$$\int_{-1/(2\Delta\tau)}^{1/(2\Delta\tau)} P_a(f)\, df = \int_{-1/(2\Delta\tau)}^{1/(2\Delta\tau)} P_A(f)\, df = \int_{-\infty}^{\infty} P_A(f)\, df = \int_{-\infty}^{\infty} P(f)\, df.$$

If we examine the relation of $P_A(f')$ to $P(f)$, where

$$0 \leq f' \leq \frac{1}{2\Delta\tau}, \quad \text{and} \quad f \geq 0,$$

we find that

$$P_A(f') = P(f') + \sum_{q=1}^{\infty}\left[P\left(\frac{q}{\Delta\tau} - f'\right) + P\left(\frac{q}{\Delta\tau} + f'\right)\right],$$

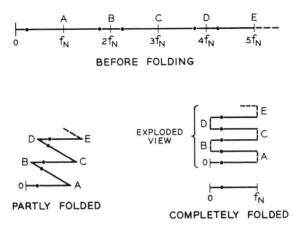

Fig. 18 — Spectrum folding.

(the right-hand side representing $P_a(f')$ for all f'). We say, therefore, that the power density at $f = (q/\Delta\tau) - f'$ and at $f = (q/\Delta\tau) + f'$ in the true spectrum $P(f)$, where $q = 1, 2, \cdots, \infty$, are *aliased* at f' in the principal part $P_A(f')$ of the aliased spectrum. Clearly, the presence of aliases in power density estimates from time series is a matter of some concern.

Aliasing is sometimes called *spectrum folding*, because the pattern by which various frequencies are aliased with one another corresponds to the result of folding up the frequency axis, as illustrated in Fig. 18. The frequency of the first fold is called the *folding* or *Nyquist frequency*. In the discussion above this was $f_N^* = 1/(2\Delta\tau)$, which usually coincides with $f_N = 1/(2\Delta t)$.

B.13 TRANSFORMATION AND WINDOWS

A general outline of a hypothetical procedure for power spectrum estimation from a uniformly spaced discrete time series of finite length, by the indirect route, is as follows:

(1) Let $X_0, X_1, \cdots, X_n$ be the time series and let Δt be the time interval between adjacent values. Compute *mean lagged products*, with lag interval $\Delta\tau = h\Delta t$, according to the formula

$$C_r = \frac{1}{n - hr} \sum_{q=0}^{q=n-hr} X_q \cdot X_{q+hr}$$

$$\left(r = 0, 1, \cdots, m \quad \text{where} \quad m \leqq \frac{n}{h} \right).$$

(In a practical procedure, this formula will have to be modified to avoid difficulties with spurious low-frequency components.)

(2) Compute *raw spectral density estimates* according to the formula

$$V_r = \Delta\tau \cdot \left[C_0 + 2 \sum_{q=1}^{q=m-1} C_q \cdot \cos \frac{qr\pi}{m} + C_m \cdot \cos r\pi \right].$$

Since V_r is symmetric in r about every integral multiple of m, it is necessary to compute it only for $r = 0, 1, \cdots, m$. The frequency corresponding to r is $r/(2m\Delta\tau)$, as shown below.

(3) Compute *refined spectral density estimates* according to the formula

$$U_r = a_{i0}V_r + \sum_{j=1}^{\infty} a_{ij}[V_{r+j} + V_{r-j}] ,$$

where the a's are the same as in Section B.5. In particular, for the third pair of lag and spectral windows described in Section B.5, we have $a_{30} = 0.54$ and $a_{31} = 0.23$, all others being zero, so that

$$U_r = 0.23 \, V_{r-1} + 0.54 \, V_r + 0.23 \, V_{r+1} .$$

(These power density estimates should of course be doubled if they are to be referred to positive frequencies only. This doubling may in fact be introduced through the mean lagged products.)

Comparing this outline with the one for the continuous case (Section B.4), it will be noted that the window is introduced by different methods. In the continuous case the window is introduced as a lag window before cosine transformation, in order to avoid convolution after transformation. In the discrete series case, since convolution is not difficult, indeed is very simple, convolution after transformation is convenient, and the lag window is shaped after the cosine transformation.

To relate the outlined procedure for the discrete series case to that for the continuous case, we note that

$$\text{ave } \{C_r\} = C(r\Delta\tau),$$

where $C(\tau)$ is the true autocovariance function. Hence,

$$\text{ave } \{V_r\} = \Delta\tau \cdot \left[C(0) + 2 \sum_{q=1}^{q=m-1} C(q\Delta\tau) \cos \frac{qr\pi}{m} + C(m\Delta\tau) \cos r\pi \right].$$

This may be expressed in the form of a Fourier transform, viz.,

$$\text{ave } \{V_r\} = \int_{-\infty}^{\infty} [\nabla_m(\tau; \Delta\tau) \cdot C(\tau)] \cdot e^{-i\omega\tau} \, d\tau ,$$

where $\nabla_m(\tau; \Delta\tau)$ is the finite Dirac comb defined in Section A.2, and

$$f = \frac{\omega}{2\pi} = \frac{r}{2m\Delta\tau}.$$

Therefore,

$$\text{ave } \{V_r\} = \left[Q_0(f; \Delta\tau) * P(f) \right]_{f=r/(2m\cdot\Delta\tau)},$$

where $P(f)$ is the true power spectrum. Hence, V_r may be regarded as an estimate of an average-over-frequency of $P(f)$ in the *aliased* neighborhood of $f = r/(2m\Delta\tau)$, with the spectral window $Q_{0A}(f) = Q_0(f; \Delta\tau)$ illustrated in Fig. 9.

Three other views of the relation of V_r to $P(f)$ may be developed from the fact that (from the end of Section A.2)

$$\nabla_m(\tau; \Delta\tau) = D_0(\tau) \cdot \nabla(\tau; \Delta\tau),$$

$$Q_0(f; \Delta\tau) = Q_0(f) * A\left(f; \frac{1}{\Delta\tau}\right),$$

whence,

$$\text{ave } \{V_r\} = \left[Q_0(f) * A\left(f; \frac{1}{\Delta\tau}\right) * P(f) \right]_{f=r/(2m\cdot\Delta\tau)}.$$

The view taken in the preceding paragraph corresponds to the substitution

$$Q_0(f) * A\left(f; \frac{1}{\Delta\tau}\right) = Q_{0A}(f).$$

A second view corresponds to the substitution

$$A\left(f; \frac{1}{\Delta\tau}\right) * P(f) = P_a(f),$$

and a third to the identity

$$A\left(f; \frac{1}{\Delta\tau}\right) * P(f) = A\left(f; \frac{1}{\Delta\tau}\right) * P_A(f).$$

The latter two correspond to the results

$$\text{ave } \{V_r\} = \left[Q_0(f) * P_a(f) \right]_{f=r/(2m\cdot\Delta\tau)},$$

$$\text{ave } \{V_r\} = \left[Q_{0A}(f) * P_A(f) \right]_{f=r/(2m\cdot\Delta\tau)}.$$

Hence, V_r may also be regarded as an estimate of an average-over-frequency of the entire *aliased power spectrum* $P_a(f)$ in the neighborhood of $f = r/(2m\Delta\tau)$, with the spectral window $Q_0(f)$ illustrated in Fig. 14, or, and usually most usefully, as an estimate of an average-over-frequency of the *principal part $P_A(f)$ of the aliased spectrum*, in the neighborhood of $f = r/(2m\Delta\tau)$, with the aliased window $Q_{0A}(f)$, which is illustrated in Fig. 9.

Finally, a fourth view corresponds to the substitution (from Section B.4)

$$Q_0(f) * P(f) = \text{ave } \{P_0(f)\}.$$

Hence, V_r may also be regarded as an aliased version of the estimated power spectrum $P_0(f)$ which would have been obtained from the continuous data in accordance with the hypothetical procedure outlined in Section B.4.

Regarded from any one of the four points of view developed above, we see that the raw spectral density estimates V_r, (assuming $\Delta\tau$ sufficiently small, so that aliasing is negligible) are in the same position as the estimated power spectrum $P_0(f)$ in the continuous case; the spectral window $Q_{0A}(f)$ is essentially $Q_0(f)$ with the same undesirable side lobes. Hence, the raw power density estimates must be refined. As in the continuous case, the refinement can be done by using a graded lag window before transformation. In the discrete series case, it may also be done by convolving the raw estimates with the a_{ij}'s of Section B.5 to obtain the refined estimates, as described under (3) in the outline.

In any event, we may write, as we did at the end of Section 4, of Part I,

$$\text{ave } \{U_r\} = \int_0^\infty H_i\left(f; \frac{r}{2m\Delta\tau}\right) \cdot P_a(f) \cdot df,$$

where

$$H_i(f; f_1) = Q_i(f + f_1) + Q_i(f - f_1).$$

Alternatively, we may write

$$\text{ave } \{U_r\} = \int_0^\infty H_{iA}\left(f; \frac{r}{2m\Delta\tau}\right) \cdot P(f) \cdot df,$$

or, more usually,

$$\text{ave } \{U_r\} = \int_0^{1/(2\cdot\Delta\tau)} H_{iA}\left(f; \frac{r}{2m\Delta\tau}\right) \cdot P_a(f) \cdot df$$

$$= \int_0^\infty H_{iA}\left(f; \frac{r}{2m\Delta\tau}\right) \cdot P_A(f) \, df,$$

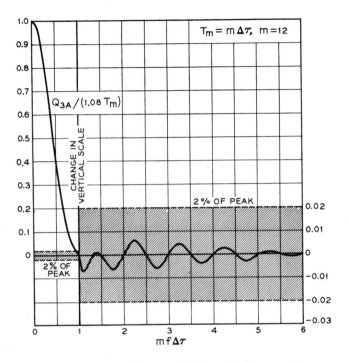

Fig. 19 — Aliased spectral window Q_{3A} for $m = 12$.

where

$$H_{iA}(f; f_1) = Q_{iA}(f + f_1) + Q_{iA}(f - f_1),$$

and

$$Q_{iA}(f) = Q_i(f) * A\left(f; \frac{1}{\Delta\tau}\right) = \sum_{q=-\infty}^{+\infty} Q_i\left(f - \frac{q}{\Delta\tau}\right).$$

In particular, $Q_{2A}(f)$ and $Q_{3A}(f)$ are illustrated in Figs. 8 and 19.

B.14 VARIABILITY AND COVARIABILITY

The analysis of variability of power density estimates from discrete time series is a repetition of Section B.6 up to and including (B-6.4) for cov $\{M(t_1, \tau_1), M(t_2, \tau_2)\}$. Beyond this point we now have to deal with summations rather than integrations with respect to t_1 and t_2.

If the range of summation in the formula for the mean lagged products, viz., $n-hr$, is replaced by an average range n', which may be regarded as the effective length of the series, and which may usually be taken as

$$n' = \text{largest integer less than } \left(n - \frac{hm}{3}\right)$$

then

$$\text{cov }\{C_r, C_s\} = \int_{-\infty}^{\infty} \cos \omega r \Delta \tau \cdot \cos \omega s \Delta \tau \cdot \Gamma_{\Delta t}(f) \cdot df,$$

where

$$\Gamma_{\Delta t}(f) = 4 \int_{-\infty}^{\infty} P(f + f') \cdot P(f - f') \cdot \left(\frac{\sin \omega' n' \Delta t}{n' \sin \omega' \Delta t}\right)^2 \cdot df'$$

(reducing to (B-6.10) and (B-6.11) as $\Delta t \to 0$ with $n'\Delta t = T'_n$). Finally, if U_r and U_s are refined power density estimates based on applying the spectral windows $Q_i(f)$ and $Q_j(f)$, respectively, to the aliased power spectrum $P_a(f)$, then

$$\text{cov }\{U_r, U_s\} = \frac{1}{4} \int_{-\infty}^{\infty} H_{iA}\left(f; \frac{r}{2m\Delta\tau}\right) \cdot H_{jA}\left(f; \frac{s}{2m\Delta\tau}\right) \cdot \Gamma_{\Delta t}(f) \cdot df.$$

In particular, of course

$$\text{var }\{U_r\} = \frac{1}{4} \int_{-\infty}^{\infty} \left[H_{iA}\left(f; \frac{r}{2m\Delta\tau}\right)\right]^2 \cdot \Gamma_{\Delta t}(f) \cdot df.$$

We see now that there is essentially no difference between power density estimation from uniformly spaced time series and power density estimation from continuous data, except for aliasing and its secondary consequences. In particular, if $P(f)$ is reasonably smooth, and if aliasing is negligible, then we may judge the stability of the power density estimates U_r for the uniformly spaced case by analogy with a chi-square variate with k degrees of freedom, where

$$k = 2\left(\frac{n\Delta t}{m\Delta\tau} - \frac{1}{3}\right), \quad r = 1, 2, \cdots, (m-1),$$

$$= \text{half as much for } r = 0, m.$$

B.15 AND 16 TRANSVERSAL FILTERING[*]

If the Z's are moving linear combinations of the X's, for example

$$Z_q = c_0 X_q + c_1 X_{q-1} + \cdots + c_k X_{q-k},$$

---

[*] See Kallmann[32] for the origin of this term.

and we take $\Delta t = 1$ for convenience, then the spectra are related by

$$P_Z(f) = P_X(f) \, | \, c_0 + c_1 e^{-i\omega} + c_2 e^{-2i\omega} + \cdots + c_k e^{-ki\omega} \, |^2$$

$$= P_X(f)[(c_0^2 + c_1^2 + c_2^2 + \cdots + c_k^2)$$

$$+ 2(c_0 c_1 + c_1 c_2 + \cdots + c_{k-1} c_k) \cos \omega$$

$$+ 2(c_0 c_2 + c_1 c_2 + \cdots + c_{k-2} c_k) \cos 2\omega + \cdots$$

$$+ 2(c_0 c_k) \cos k\omega],$$

where $\omega = 2\pi f$. The first equality arises by considering $X_q \equiv e^{iq\omega}$, and the second involves the sequence of coefficients

$$b_{-k} = c_0 c_k,$$

$$b_{-k+1} = c_0 c_{k-1} + c_1 c_k,$$

$$\vdots$$

$$b_{-1} = c_0 c_1 + c_1 c_2 + \cdots + c_{k-1} c_k,$$

$$b_0 = c_0^2 + c_1^2 + \cdots + c_k^2,$$

$$b_1 = c_1 c_0 + c_2 c_1 + \cdots + c_k c_{k-1},$$

$$b_2 = c_2 c_0 + c_3 c_1 + \cdots + c_k c_{k-2},$$

$$\vdots$$

$$b_{k-1} = c_{k-1} c_0 + c_k c_1,$$

$$b_k = c_k c_0,$$

which represent the convolution of the sequence c_0, c_1, $\cdots$, c_k with itself.

As a filter, the moving linear combination is characterized by

$$| \, Y(f) \, |^2 = b_0 + 2 \, b_1 \cos \omega + 2 \, b_2 \cos 2\omega + \cdots + 2 \, b_k \cos k\omega,$$

which is never negative (as the square of an absolute value). Since we can write $\cos j\omega$ as a polynomial in $\cos \omega$ of degree j, we know that:

(1) $| \, Y(f) \, |^2$ can be written as a polynomial of degree k in $\cos \omega$,

(2) for $-1 \le \cos \omega \le 1$, it is not negative.

Any such polynomial can be realized as a moving linear combination (in several ways, see Wold[33]). The simplest way to see this is to factor the given polynomial into linear and quadratic factors. By appropriate choice of signs these factors will satisfy (2), and, if each corresponds to a real moving linear combination, the moving linear combination obtained by applying them successively will correspond to the given polynomial.

Any such linear factor takes the form

$$| Y_i(f) |^2 = A^2(1 + a \cos \omega)$$

with $| a | \leqq 1$, which may be realized by $k = 1$ and

$$c_0, c_1 = \tfrac{1}{2}A(\sqrt{1 + a} \pm \sqrt{1 - a}).$$

Any such quadratic factor takes the form

$$| Y_j(f) |^2 = A^2(1 + a_1 \cos \omega + a_2 \cos^2 \omega).$$

The condition for this to be non-negative for $| \cos \omega | \leqq 1$ is

$$| a_1 | \leqq 1 + a_2,$$

and if $a_2 \geqq 1$, so that the first condition forces an internal extremum on $[-1, +1]$, we must also have $a_1^2 \leqq 4a_2$. We may write

$$| Y_j (f) |^2 = A^2 \left(1 + \frac{a_2}{2} + a_1 \cos \omega + \frac{a_2}{2} \cos 2\omega\right),$$

which is realized by a three-point moving linear combination with

$$c_0, c_2 = \frac{A}{2\sqrt{2}} (\sqrt{1 + a_2 + \sqrt{}} \pm \sqrt{1 - a_2 + \sqrt{}}),$$

$$\sqrt{} = \sqrt{(1 + a_2)^2 - a_1^2},$$

$$c_1 = \frac{A}{2} (\sqrt{1 + a_2 + a_1} - \sqrt{1 + a_2 - a_1}),$$

the conditions stated above ensuring that all radicals are real.

These two cases — linear and quadratic factors — not only prove that every polynomial non-negative on $(-1, +1)$ can be obtained, but provide at least one way to find a moving linear combination with an assigned polynomial.

One special case deserves record. If we require a simple moving linear combination with $Y(f_0) = 0$, we may use

$$c_0 = \frac{1}{2 \cos \omega_0},$$

$$c_1 = -1,$$

$$c_2 = \frac{1}{2 \cos \omega_0},$$

which has $| Y(f) |^2 = [1 - (\cos \omega / \cos \omega_0)]^2.$

Now that methods are available for finding moving linear combinations whose spectral windows are prescribed polynomials in cos ω, some attention should be given to approximating an arbitrary desired response by polynomials. If a roughly "equal ripple" approximation (where the local maxima deviations of approximation from desired are roughly equal) is desired, then the techniques described in the next paragraph will be quite effective. If, as seems likely, however, we desire the *fractional* error

$$\frac{\text{approximation} - \text{desired}}{\text{desired}}$$

to have roughly equal ripples, then no specific method seems to be available. All we can suggest is the following procedure: (i) find a roughly equal ripple approximation, (ii) find the zeros of its error, (iii) squeeze these zeros together where smaller ripples are desired, and open them out where larger ripples can be permitted, keeping as much of the same general pattern of distances between zeros as possible, (iv) construct a new polynomial with these points for the zeros of its error, (v) adjust the result slightly, if necessary, to make it non-negative. (This procedure sounds quite plausible, but the reader should be warned that we do not know how to be more specific about "keeping as much of the same general pattern of distances as possible". However, we expect the procedure to work in many hands.)

The construction of the roughly "equal ripple" approximation can proceed in many ways. In almost every case, one should begin by calculating values of the desired response at values of cos ω equally spaced from cos ω = +1 to cos ω = −1. The semi-classical approach (DeLury,[34] Fisher-Yates,[35] or Milne[36]) would be to fit orthogonal polynomials to these equi-spaced points. The results would be least-square fits, but might be far from equal-ripple. The process of "economization" (Lanczos,[37] Lanczos,[38] streamlined by Minnick[39]) will allow us to take an over-fitted least-square fit and back up to an "equal-ripple" polynomial of lower order. However, the direct attack, based on central differences of the desired response (at equi-spaced values of cos ω) as proposed by Miller[40] seems likely to be shorter, even allowing for the expansion of the Čebyšev or Chebyshёv polynomial series into single polynomials.

One further set of considerations remains which is sometimes important. These relate to the starting up of a moving linear combination. If Z_q involves X_q back to X_{q-k}, then there will be k less Z's than X's, and nothing can be done about this. Any transversal filtration causes the loss of some data, and if the filter characteristic is complicated (as a polynomial in cos ω) the loss will have to be correspondingly great. This is usually unimportant, but, with very short pieces of record, might become crucial.

In Section 15 we remarked that an autoregressive relation, e.g.,

$$X_q = c_0 Z_q + c_1 Z_{q-1} + \cdots + c_k Z_{q-k},$$

between X's and Z's enabled us to obtain the reciprocal of any suitably non-negative cosine polynomial as the ratio of the spectrum of Z to the spectrum of X. There are different ways of looking at the situation which make this statement true, not true, or partly true. If we have all the X's back to $q = -\infty$, we can calculate the corresponding Z's and it is true. If we have only a finite number of X's, as always in practice, then we have to start the calculation up in some other way, perhaps like this

$$X_0 = c_0 Z_0,$$

$$X_1 = c_0 Z_1 + c_1 Z_0,$$

$$\vdots$$

$$X_{k-1} = c_0 Z_{k-1} + c_1 Z_{k-2} + \cdots + c_{k-1} Z_0,$$

$$X_k = c_0 Z_k + c_1 Z_{k-1} + \cdots + c_k Z_0.$$

As a consequence, we will have introduced an initial transient into the form of the autoregressive transformation so that our Z's are never related to the X's in the way we supposed. In this sense the statement is untrue. In many cases, however, this initial transient dies out quite quickly, and if we discard enough initial Z's, perhaps $2k$ to $4k$ of them, we can regard the reciprocal of the cosine polynomial as a satisfactory approximation. In this sense the statement is partly true.

In theory, an autoregressive scheme corresponds to an infinitely long moving linear combination. In practice it corresponds to a sequence of changing moving linear combinations of finite but increasing length which approximate the infinitely long one. Sometimes the approximation is quite good enough. (Perhaps the main advantage to the autoregressive scheme is its likely reduction in arithmetic — a few autoregressive coefficients corresponding to a long moving linear combination.)

B.17 SMOOTHING AND DECIMATION PROCEDURES

We now study the effects of applying, successively and in any order, simple smoothing and decimation. The basic operations are taking equally weighted means of ℓ consecutive values, and discarding all but every jth value. Simple sums are usually more convenient than means, and, since the results differ only by a fixed constant factor, lead to the same spectra except for a constant. We define, then, as our basic operations, S_ℓ and F_j, where (for definiteness)

$$Y = S_\ell X$$

means

$$Y_i = X_i + X_{i+1} + \cdots + X_{i+\ell-1}, \qquad (\ell \text{ terms}),$$

while

$$Z = F_j X$$

means

$$Z_i = X_{1+(i-1)j}.$$

It will also be convenient in dealing with the algebra of these operations, to use an operator for simple summing at wider spacings. We therefore define $S_\ell^{(h)}$ by taking

$$W = S_\ell^{(h)} X$$

to mean

$$W_i = X_i + X_{i+h} + \cdots + X_{i+(\ell-1)h}, \qquad (\ell \text{ terms}).$$

It is immediately clear that

$$S_\ell^{(h)} S_h = S_{\ell h} \qquad (\text{B-17.1})$$

both sides corresponding to forming sums of ℓh consecutive X_t's.

Now, consider the equation

$$Z = S_\ell F_j X,$$

which means that if

$$Y = F_j X,$$

then

$$Z = S_\ell Y.$$

These, in turn, mean that

$$Y_i = X_{1+(i-1)j},$$

and

$$Z_i = Y_i + Y_{i+1} + \cdots + Y_{i+\ell-1}.$$

Hence,

$$Z_i = X_g + X_{g+j} + \cdots + X_{g+(\ell-1)j},$$

where $g = 1 + (i - 1)j$. Now, since

$$W = S_\ell^{(j)} X$$

means

$$W_g = X_g + X_{g+j} + \cdots + X_{g+(\ell-1)j},$$

we have

$$Z_i = W_{1+(i-1)j},$$

which corresponds to

$$Z = F_j W.$$

Thus, we have shown that

$$S_\ell F_j = F_j S_\ell^{(j)}. \tag{B-17.2}$$

Similarly, we find that

$$S_\ell^{(h)} F_j = F_j S_\ell^{(jh)}. \tag{B-17.3}$$

The order in which F_j and S_ℓ or $S_\ell^{(h)}$ are applied is thus important. On the other hand, it is easy to show that

$$S_\ell S_h = S_h S_\ell \neq S_{\ell h}, \tag{B-17.4}$$

and

$$F_j F_h = F_h F_j = F_{jh}. \tag{B-17.5}$$

Thus, sample reductions are

$$S_2 F_4 S_4 = F_4 S_2^{(4)} S_4 = F_4 S_8,$$

and

$$F_2 S_4 F_2 S_2 = F_2 F_2 S_4^{(2)} S_2 = F_4 S_8.$$

It may be noted also that $F_\ell S_\ell$ corresponds to "summing in (barely) non-overlapping blocks of ℓ terms".

A reason for these differing relations is easily found. The changes in the spectrum due to S_ℓ and F_j are quite different in character. S_ℓ multiplies spectra by the power transfer function

$$\mid 1 + e^{-i\omega\Delta t} + \cdots + e^{-i(\ell-1)\omega\Delta t} \mid^2 = \left(\frac{\sin \dfrac{\ell\omega\Delta t}{2}}{\sin \dfrac{\omega\Delta t}{2}} \right)^2,$$

while $S_\ell^{(h)}$ multiplies spectra by

$$\left(\frac{\sin \dfrac{\ell h\omega\Delta t}{2}}{\sin \dfrac{h\omega\Delta t}{2}}\right)^2 .$$

On the other hand, F_j changes the folding frequency, f_N, to a new value $(1/j)^{\text{th}}$ as large as before, namely

$$f_N' = f_N/j,$$

and aliases together the old principal aliases in sets of j. The j old principal aliases which have f as their new principal alias are

$$f,\ 2f_N' - f,\ 2f_N' + f,\ 4f_N' - f,\ 4f_N' + f,\ \cdots\ (j\ \text{terms}).$$

Our combined operations will, in general, change both amplitudes and principal aliases. If

$$P_{+A}(f) = T(f)P_A(f) + T(2f_N' - f)P_A(2f_N' - f)$$
$$+ T(2f_N' + f)P_A(2f_N' + f) + \cdots$$

where $P_{+A}(f)$ is the new aliased spectrum, $P_A(f)$ is the old aliased spectrum, f_N' is the new folding (Nyquist) frequency, and the summation continues as long as the arguments lie below the old folding frequency, then we call $T(f)$, for any f up to the old folding frequency, the *transmission*. (If there is no new aliasing, the transmission is merely the power transfer function.) The ratio of transmission for a desired frequency to the transmission for an undesired alias of that frequency, such as $T(f)/T(2f_N' - f)$ for example, will be called the *protection ratio*.

Fig. 20 illustrates some simple cases. Curve (b) is the transmission of S_3, F_2S_3, or, indeed any F_jS_3. All of these are given by

$$\left(\frac{\sin \dfrac{3\omega\Delta t}{2}}{\sin \dfrac{\omega\Delta t}{2}}\right)^2 .$$

Scale (c) illustrates the reduced range of the principal aliases for F_2, F_2S_3, or, indeed any $S_jF_2S_k$. Curve (d) shows the power transfer function of S_3, regarded as following F_2. In terms of the respective folding frequencies, this curve duplicates curve (b). On an absolute frequency scale it is different. Curve (e) presents the transmission corresponding to $S_3F_2 = F_2S_3^{(2)}$ which aliases into curve (d).

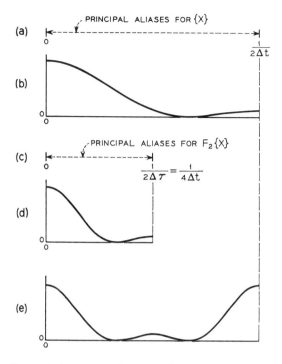

Fig. 20 — Transmission curves for a simple example of smoothing and decimation.

The transmission (= power transfer function) of S_8 is shown in Fig. 21.

The smoothing operation described in Section 17 was, in our present notation, $F_j S_k$ where $j = k/4$, which equals $S_4 F_j S_j$. In the case $j = 2$, we have $F_2 S_8 = S_4 F_2 S_2$, and the highlights of transmission and folding pattern are as follows, where the frequency scale has been chosen to make the original folding frequency $= 8$.

Original freq.	0	1	2	3	4	5	6	7	8
Aliased freq.	0	1	2	3	4	3	2	1	0
Transmission	64	26	zero	3.2	zero	1.4	zero	1.0	zero

The protection provided for frequency 1 against aliasing from frequency 7 is only in the ratio of 26 to 1. If the spectrum falls toward higher frequencies, this may be enough, but adequacy is far from certain.

Double use of S_8 before selecting every second observation would square this protection ratio, giving a ratio near 700. If we are to sum by groups twice, however, we can do better than to use the same length

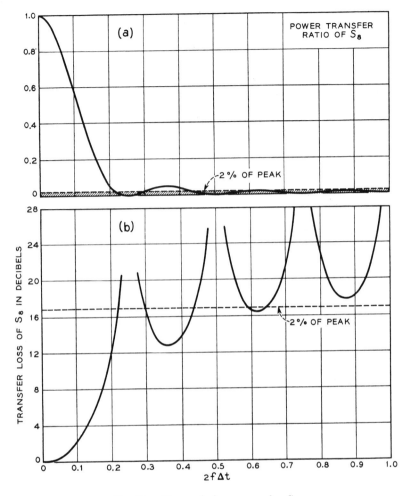

Fig. 21 — Transmission curves for S_8.

of group twice. By using slightly different lengths, we can spread out the zeros of the transmission curve, and tend to hold the right-hand end of the transmission curve nearer the origin. Table V shows the transmission and folding behavior of $F_3S_7S_8$, with the original folding frequency taken as 18. Over the lower half of the folded spectrum, this choice yields protection (against aliasing) by a ratio of about 3,000, which should suffice under even moderately extreme circumstances.

If we do not wish to fold quite so far (or in multiples of 2) then $F_2S_5S_6$ gives a protection ratio of 1,500 or better over the lower half of the folded spectrum.

TABLE V — TRANSMISSION AND FOLDING BEHAVIOR OF $F_3S_7S_8$
OR $F_3S_8S_7$ (ORIGINAL FOLDING FREQUENCY $= 18$)

Original Frequency	Folded Frequency	Relative Transmission
0	0	3136
1	1	2356
2	2	946
3	3	156
4	4	3.5
5	5	0.03
6	6	3.0
7	5	7.4
8	4	2.4
9	3	zero
10	2	0.04
11	1	0.39
12	0	1.0
13	1	0.17
14	2	0.09
15	3	0.06
16	4	0.12
17	5	0.28
18	6	zero

So long as we are satisfied with such, only moderately large, protection ratios, and with a substantial fall-off of transmission over the useful part of the spectrum, such repeated *unweighted* summing-or-averagings by groups are likely to be most desirable from the point of view of machine computation. If requirements on the smoothing-and-decimation operation are more stringent, smoothing with a suitably chosen set of graded weights is likely to be required.

B.18 MODIFIED PILOT ESTIMATION, CASCADE ESTIMATION

The addition-and-subtraction calculation discussed in Section 18 (i) yields only one estimate per octave, (ii) is unduly sensitive to trends, (iii) involves windows which are broad even for a one-per-octave spacing, and (iv) is deliberately wasteful of data in the interest of computational simplicity. There may well be a desire to correct any or all of these.

The δ operator, in a notation extending that of the last section, takes the form $\delta = F_2D$, where F_2 is the operation of dropping every alternate value and D is the operation of differencing adjoining values. We may replace δ by D in the computing routine, keeping σ the same, with the following effects: (i) loss of symmetry in the procedure, (ii) introduction of numerical factors like 8/15 (to be applied to the sums of squared differences), (iii) near doubling of number of differences available for squaring and summing. If not enough data is available to give adequate stability to the results of the standard pilot method, then this modification may be worth while.

If we denote the operation of omitting the *other* half of the values by F_2' and introduce $\sigma' = F_2'S_2$ as well as $\sigma = F_2S_2$, then, by using D, σ and σ' successively, we may obtain even more differences for squaring and summing. Table VI shows a convenient pattern of computation, in which sums and differences are *not* located in lines near the lines from which they are derived. Table IX, below, (in Section B.28) provides a numerical example.

TABLE VI — COMPACT CALCULATION ARRANGEMENT FOR COMPLETE
VERSION OF PILOT ESTIMATION PROCEDURE

X	DX	$\left\{{\sigma \atop \sigma'}\right\}X$	$D\left\{{\sigma \atop \sigma'}\right\}X$	$\left\{{\sigma \atop \sigma'}\right\}\left\{{\sigma \atop \sigma'}\right\}X$	$D\left\{{\sigma \atop \sigma'}\right\}\left\{{\sigma \atop \sigma'}\right\}X$	$\left\{{\sigma \atop \sigma'}\right\}\left\{{\sigma \atop \sigma'}\right\}\left\{{\sigma \atop \sigma'}\right\}X$
X_1		SX_1		SSX_1		$SSSX_1$
	DX_1		DSX_1		$DSSX_1$	
X_2		SX_3		SSX_5		
	DX_2		DSX_3			
X_3		SX_5				
	DX_3		DSX_5			
X_4		SX_7		SSX_3		
	DX_4					
X_5						
	DX_5					
X_6		SX_2		SSX_2		
	DX_6		DSX_2			
X_7		SX_4				
	DX_7		DSX_4			
X_8		SX_6		SSX_4		

$$DX_q = X_{q+1} - X_q, \qquad SSX_q = (SX_{q+2}) + (SX_q)$$
$$SX_q = X_{q+1} + X_q, \qquad DSSX_q = (SSX_{q+4}) - (SSX_q)$$
$$DSX_q = (SX_{q+2}) - (SX_q) \qquad SSSX_q = (SSX_{q+4}) + (SSX_q)$$

Two methods are available to cope with trend difficulties. The original series may be differenced, and the differenced series subjected to pilot estimation. (The main disadvantage being some loss in accuracy, etc., at very low frequencies.) Or each column of differences may be "corrected for its mean" adjusting the corresponding divisor from k to $k - 1$. (This is only recommended when the first modification is in use and the column contains all D's, not merely the corresponding δ's. Instead of

$$\frac{k+1}{2k} \sum_1^k (D_q)^2$$

then, the corrected sum, corresponding to $(k + 1)/2$ squares, becomes

$$\frac{k+1}{2(k-1)} \left[\sum_1^k (D_q)^2 - \frac{1}{k} \left(\sum_1^k D_q \right)^2 \right].$$

In view of the fact that $\sum D_q$ should equal the last value in the preceding column minus the first value there, a convenient check on the D_q exists.

If, as may not be too unlikely, none of these modifications suffices, as will surely be the case if more than one estimate per octave is required, something better than the modified pilot estimation method, ·without going all the way to the detailed method's equi-spacing in frequency, may be desired. Such an intermediate method should give roughly constant spacing on a logarithmic scale, and provide reasonably clean windows, with about a 100-to-1 ratio between major lobes and minor lobes. It would be useful for high quality pilot estimation, and might sometimes suffice for the complete analysis.

Easy calculation and a roughly logarithmic scale both favor a cascade process, in each of whose cycles some computation is carried out on given values, and then half as many values are computed ready for the next cycle. We may expect, then, that it will be possible to think of each cycle as having three phases:

(a) computation of estimates
(b) smoothing of given values
(c) deletion of alternate smoothed values.

Our main attention needs to be given to the last two phases, since the first is a branch which can be changed rather freely.

We are concerned, therefore, with smoothing procedures to precede halving of the folding frequency. We have, then, to choose two frequencies averaging to the new folding frequency such that we plan to make estimates based on the smoothed (and halved) values up to the lower of these, while effectively eliminating frequencies above the higher. If the folding frequency at the start of the present cycle is f_0, these two frequencies can be written as αf_0 and $(1 - \alpha)f_0$. In the next cycle, then, we anticipate estimation up to αf_0. In the present cycle, we anticipate estimation up to $2\alpha f_0$ and must consequently cover the octave from αf_0 to $2\alpha f_0$.

In the choice of α we must balance two considerations of computing effort. For a given number of lags (a given number of multiplications in forming mean lagged products per cycle) our estimates are spaced a fixed fraction of f_0. The larger α and $2\alpha f_0$, the more of these we may use. Contrariwise, the smaller α, the easier it is, in terms of computational effort, to provide smoothing which suppresses the whole interval from $2\alpha f_0$ to f_0 by a factor of about 100 in comparison with what it does to any frequency between 0 and αf_0.

Trial suggests that a value of α near $\frac{1}{3}$ is reasonable. We wish, therefore, a smoothing which suppresses frequencies between $(\frac{2}{3})f_0$ and f_0.

Two elementary smoothings with zeros in this range are running means-or-sums of 2 (with a zero at f_0) and running means-or-sums of 3 (with a zero at $(\frac{2}{3})f_0$). If, for convenience, we use sums, these operations multiply the spectrum by, respectively,

$$2 + 2 \cos \pi f/f_0,$$

and

$$3 + 4 \cos \pi f/f_0 + 2 \cos 2\pi f/f_0.$$

If both are applied, the factor takes on the following values:

$f/f_0 =$	0.0	0.3	$\frac{1}{3}$	0.5	0.6	0.62	$\frac{2}{3}$	0.8	0.82	$\frac{5}{6}$	1
factor =	36	15	12	2	0.23	0.08	0	0.146	0.147	0.143	0

The result is a protective factor (against the effects of aliasing on halving) of just over 100 for $\alpha = 0.3$ and of about 80 for $\alpha = \frac{1}{3}$. This should be entirely satisfactory for most pilot purposes. (If further protection is needed, $Z_q = 0.6X_{q-1} + X_q + 0.6X_{q+1}$, which has zero transmission at $f/f_0 = 0.815$, could also be applied.)

If we wish to reduce the number of additions, we might smooth by threes, and then combine smoothing by twos and halving, e.g.,

$$X_q^* = X_{q-1} + X_q + X_{q+1},$$
$$Z_q = X_{2q}^* + X_{2q+1}^*,$$

which requires 2.5 additions per X-value. If we rearrange, however, to

$$\tilde{X}_{2q} = X_{2q} + X_{2q+1},$$
$$2\tilde{X}_{2q} = \tilde{X}_{2q} + \tilde{X}_{2q},$$
$$Z_q = X_{2q-1} + 2\tilde{X}_{2q} + X_{2q+2},$$

we find only 2 additions per X-value. Thus this type of smoothing and halving is computationally quite simple.

If we use this smoothing-and-halving process cycle after cycle we must, after obtaining our spectral estimates for the series actually processed in a given cycle, adjust them for the effects of all the smoothings in preceeding cycles. Near zero frequency this factor is $(36)^{-d}$ for d previous smoothings. At a few selected frequencies the factors are as follows:

$f/f_0 =$	0	0.075	0.15	0.225	0.30
$j = 1$	$(36)^{-1}$	$1.05(36)^{-1}$	$1.23(36)^{-1}$	$1.60(36)^{-1}$	$2.4(36)^{-1}$
$j = 2$	$(36)^{-2}$	$1.06(36)^{-2}$	$1.29(36)^{-2}$	$1.81(36)^{-2}$	$2.95(36)^{-2}$
$j = 3$	$(36)^{-3}$	$1.07(36)^{-3}$	$1.31(36)^{-3}$	$1.86(36)^{-3}$	$3.10(36)^{-3}$
$j = 4$	$(36)^{-4}$	$1.07(36)^{-4}$	$1.31(36)^{-4}$	$1.87(36)^{-4}$	$3.16(36)^{-4}$

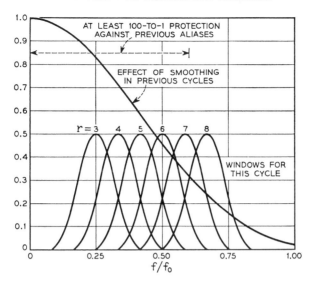

Fig. 22 — Windows for $m = 12$, and smoothing from previous cycles.

When an ordinary calculation with $m = 12$, for example, and hanning is used at each cycle, the spectral windows of the more relevant estimates and the effect of smoothing in previous cycles appear as in Fig. 22. Clearly the estimates for $r = 3, 4, 5$ and probably 6 will be quite usable. The estimate for $r = 7$ may well be usable, but its window extends up to a point where protection against aliasing is beginning to be much reduced. Computation for all r from 0 to 12 is probably worthwhile, lower values of r providing rough checks for later cycles and higher values indicating the extent of the danger from aliasing (to estimates of the next cycle) during the smooth-and-halve phase of the present cycle.

The resulting spectral windows are shown for parts of 4 cycles in Fig. 23. The windows 6, 5, 4, 3 $\sim$ 6′, 5′, 4′, 3′ $\sim$ 6″, 5″, 4″, 3″ $\sim$ 6‴, 5‴, 4‴, 3‴, and so forth, will give a fairly effective set of coarsely spaced spectral estimates.

B.19 REJECTION NEAR ZERO FREQUENCY

We come now to the details of compensation, on the average, for non-zero but constant averages or for averages changing linearly with time. (As a convenient shorthand we will refer to these as "constant" and "linear" trends.) The X_h, on a sample of which our calculations are to be based, can each usefully be regarded as a sum of two terms: (i) a fixed (but unknown) trend and (ii) a sample from a stationary ensemble with average zero. These terms are added together and are statistically independent (since one is fixed).

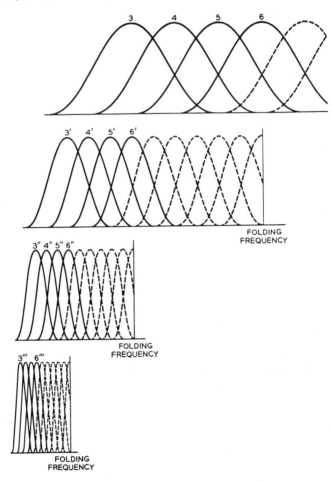

Fig. 23 — Spectral windows in successive cycles (smoothing corrected for).

Consider any quadratic expression in the X_h. If we introduce the two terms for each X_h, we may write the quadratic expression as a sum of three parts: (i) a quadratic in the trend (fixed, of course), (ii) an expression linear both in the trend and in the stationary fluctuations (and hence of average value zero), and (iii) a quadratic in the stationary fluctuations. The first and last of these three parts are, of course, just what would have arisen had the trend alone or, respectively, the fluctuations alone been present. Hence, since the average value of the middle part vanishes, the average value of the whole quadratic is the sum of the average value for the trend alone and the average value for the fluctuations alone. To study the efficacy with which a quadratic expression rejects a trend in the presence of fluctuation, *we have only to study its behavior in the presence of trend alone.* After this, we shall want to study the behavior for fluctuations alone of those expressions which satisfactorily reject trend.

(The variance of the quadratic expression will be determined from the middle and last parts. If third moments of the fluctuations vanish, as will be the case for Gaussian ensembles, the contributions will come from these parts separately, the covariance between parts vanishing.)

A few formulas will be useful in discussing possible rejection techniques. These are conveniently derived on the basis of $n = 2k + 1$ available equi-spaced values extending from $-k$ through 0 to $+k$. (The indexing of the values is only a convenience, and cannot affect any essential result. The limitation to an odd number of points is essentially a convenience also — we shall replace $2k + 1$ by an unrestricted n rather freely.)

The first formulas relate to constants and linear trends, and are as follows:

$$\text{ave } \{C_r \mid X_t \equiv 1\} = \frac{1}{2k + 1 - r} \sum_{-k}^{k-r} (1)(1) = 1,$$

$$\text{ave } \{C_r \mid X_t \equiv t\} = \frac{1}{2k + 1 - r} \sum_{-k}^{k-r} h(h + r)$$

$$= \frac{1}{6} (2k(k + 1) - r(2k + 1) - r^2),$$

$$= \frac{1}{12} n^2 K_r,$$

where

$$K_r = 1 - \frac{1}{n^2} - 2 \frac{r}{n} - 2 \frac{r^2}{n^2} \approx 1,$$

whence,

$$\text{ave } \{C_r \mid X_t \equiv \alpha + \beta t\} = \frac{1}{2k + 1 - r} \sum_{-k}^{k-r} (\alpha + \beta h)(\alpha + \beta h + \beta r)$$

$$= \alpha^2 + \frac{\beta^2}{12} n^2 K_r,$$

while

$$\frac{1}{c + d + 1} \sum_{-c}^{d} (\alpha + \beta h) = \alpha + \beta \frac{d - c}{2},$$

$$\frac{1}{a + b + 1} \sum_{-a}^{b} \frac{1}{c + d + 1} \sum_{-c}^{d} (\alpha + \beta h)(\alpha + \beta g)$$

$$= \alpha^2 + \alpha\beta \left(\frac{d - c + b - a}{2} \right) + \beta^2 \frac{(d - c)(b - a)}{4}.$$

From the first two formulas, as combined in the fourth, we learn what dependence on α and β is required if we are to reject both constants and linear trends for any fixed value of r. The fifth formula, as developed into the last, shows that this cannot be done by subtracting the product of any two simple, equally weighted means of the X's.

Suppose now that $k = 3j - 2$, so that the means of the lower one-third of all, of the upper one third of all, and of all X's are, respectively

$$\overline{X^-} = \frac{1}{2j - 1} \sum_{-k}^{-i} X_h,$$

$$\overline{X^+} = \frac{1}{2j - 1} \sum_{i}^{k} X_h,$$

$$\bar{X} = \frac{1}{6j - 3} \sum_{-k}^{k} X_h,$$

and so that, when $X_t = \alpha + \beta t$, we have

$$\overline{X^-} = \alpha - \beta \left(\frac{k + j}{2} \right) = \alpha - (2j - 1)\beta = \alpha - \frac{n}{3} \beta,$$

$$\overline{X^+} = \alpha + \beta \left(\frac{k + j}{2} \right) = \alpha + (2j - 1)\beta = \alpha + \frac{n}{3} \beta,$$

$$\bar{X} = \alpha.$$

Thus if we wish to reject a constant we may choose

$$E_{0r} = (\bar{X})^2 \text{ (independent of } r)$$

and have

$$\text{ave } \{C_r \mid X_t \equiv \alpha\} \equiv \text{ave } \{E_{0r} \mid X_t \equiv \alpha\}.$$

This choice, however, produces no rejection of $X_t \equiv \beta t$ at all.

If we wish to reject either a constant or a linear trend or both in a simple way, we may choose

$$E_{1r} = \bar{X}^2 + \frac{3}{16} K_r \cdot (\overline{X^+} - \overline{X^-})^2$$

for which

$$\text{ave } \{C_r \mid X_t \equiv \alpha + \beta t\} \equiv \text{ave } \{E_{1r} \mid X_t \equiv \alpha + \beta t\}.$$

An alternate calculation, nearly equivalent to the use of E_{0r} is illustrated by

$$\frac{1}{2k + 1 - r} \sum_{-k}^{k-r} (X_h - \bar{X})(X_{h+r} - \bar{X})$$

$$= \frac{1}{2k + 1 - r} \sum_{-k}^{k-r} X_h X_{h+r} - \bar{X}^2$$

$$- \bar{X} \left(\frac{r\bar{X} - \sum_{k-r+1}^{k} X_h}{2k + 1 - r} + \frac{r\bar{X} - \sum_{-k}^{-k+r-1} X_h}{2k + 1 - r} \right)$$

$$= \frac{1}{2k + 1 - r} \sum_{-k}^{k-r} X_h X_{h+r} - \bar{X}^2 - \bar{X} Q_r,$$

where

$$Q_r = \frac{1}{2k + 1 - r} \left[2r\bar{X} - \sum_{k-r+1}^{k} X_h - \sum_{-k}^{-k+r-1} X_h \right].$$

Thus subtracting the mean of all the observations from each observation before forming the C_r's is equivalent to using

$$E_{2r} = \bar{X}^2 + \bar{X} Q_r.$$

Similarly, fitting and subtracting the elementary-least-squares straight line corresponds to

$$\hat{\beta} = 3 \frac{\sum_{-k}^{k} h X_h}{k(k + 1)(2k + 1)} = \frac{12}{n(n^2 - 1)} \sum_{-k}^{k} h X_h,$$

and

$$\frac{1}{2k + 1 - r} \sum_{-k}^{k-r} (X_h - \bar{X} - \hat{\beta}h)(X_{h+r} - \bar{X} - \hat{\beta}(h + r))$$

$$= \frac{1}{2k + 1 - r} \sum_{-k}^{k-r} X_h X_{h+r} - \bar{X}^2 - \bar{X} Q_r$$

$$+ \frac{\hat{\beta}}{n - r} \sum_{k-r+1}^{k} (h + r) \cdot (X_h - X_{-h}) - \frac{n^2 \hat{\beta}^2}{12} \left[\frac{2(n^2 - 1)}{n(n - r)} - K_r \right].$$

Thus linear fitting corresponds to the use of

$$E_{3r} = \bar{X}^2 + \bar{X}Q_r - \frac{\hat{\beta}}{n-r} \sum_{k-r+1}^{k} (h+r)\cdot(X_h - X_{-h})$$
$$+ \frac{n^2\hat{\beta}^2}{12}\left[\frac{2(n^2-1)}{n(n-r)} - K_r\right],$$

and if we wish to simplify computation by neglecting "end corrections", we may also consider

$$E^*{}_{3r} = \bar{X}^2 + \bar{X}Q_r + \frac{n^2\hat{\beta}^2}{12} K_r.$$

Another version can be obtained by recalling the usual formulation of a sample estimate of a covariance between two unlagged variates. If we write

$$\overline{X^-}_{(r)} = \frac{1}{2k+1-r} \sum_{-k}^{k-r} X_h$$

$$\overline{X^+}_{(r)} = \frac{1}{2k+1-r} \sum_{-k}^{k-r} X_{h+r}$$

then we are led to use the expression

$$\frac{1}{2k+1-r} \sum_{-k}^{k-r} (X_h - \overline{X^-}_{(r)})(X_{h+r} - \overline{X^+}_{(r)})$$
$$= \frac{1}{2k+1-r} \sum_{-k}^{k-r} X_h X_{h+r} - \overline{X^-}_{(r)}\overline{X^+}_{(r)}$$

which corresponds to the use of

$$E_{4r} = \overline{X^-}_{(r)}\overline{X^+}_{(r)}$$

or, in the case of a linear trend, perhaps, to

$$E_{5r} = \overline{X^-}_{(r)}\overline{X^+}_{(r)} + \frac{n^2\hat{\beta}^2}{12}\left[K_r + \frac{3r^2}{n^2}\right].$$

(The E's corresponding to fitting a separate straight line to $X_{-k}, \cdots, X_{k-r}$ and to $X_{-k+r}, \cdots, X_k$ do not seem worth writing down.)

Having now obtained a collection of quadratic expressions which, when subtracted from the corresponding C_r, remove the average effect of trend, *we have now to learn what effect these subtractions will have on the contribution of the fluctuations to the average values of the corresponding quantities.* As noted at the beginning of this section, it suffices to consider pure fluctuations, so that we need only study the spectral windows corresponding to the E_{kr}, and to their finite cosine transforms.

If we consider white noise (of unit variance) for a moment, the X_h become independent with average zero and unit variance and we can calculate the average values of the various E_{kr} in an elementary way. This is of interest, because these average values for white noise are exactly proportional to the integrals of the corresponding spectral windows. We find the following results:

Quadratic expression	Average value for unit white noise	Approximation to this
E_{0r}	$\dfrac{1}{n}$	$\dfrac{1}{n}$
E_{1r}	$\dfrac{1}{n}\left(1 + \dfrac{9K_r}{8}\right)$	$\sim \dfrac{2}{n}$
E_{2r}	$\dfrac{1}{n}$	$\dfrac{1}{n}$
E_{3r}^{*}	$\dfrac{1}{n}\left(1 + \dfrac{n^2}{n^2-1}K_r\right)$	$\sim \dfrac{2}{n}$
E_{4r}	$\dfrac{n-2r}{(n-r)^2}$	$\sim \dfrac{1}{n}$
E_{5r}	$\dfrac{n-2r}{(n-r)^2} + \dfrac{n}{n^2-1}\left(K_r + \dfrac{3r^2}{n^2}\right)$	$\sim \dfrac{2}{n}$

The quadratic expressions which eliminate the average effect of a constant trend (one constant) have spectral windows integrating approximately proportional to $1/n$, while those which compensate for general linear trends (two constants) integrate approximately proportional to $2/n$. This is simple, and seems straightforward.

It is possible, however, to eliminate the average effect of a centered linear trend (one with $\alpha = 0$) by subtracting a windowless quadratic expression, one whose average vanishes for all *stationary* ensembles. We need only consider

$$X_{-2}X_0 - 2X_{-1}X_1 + X_0X_2$$

whose average value clearly vanishes under stationarity, but whose value when $X_t \equiv \alpha + \beta t$ is $2\beta^2$, to see that this is so. (Cursory inquiry into variability suggests that such use of windowless quadratics may increase the variability of the finally resulting spectral density estimates.)

Leaving such possibilities aside, let us compare the E_{kr} for $k = 0, 1, 2, 3, 4, 5$. The corresponding spectral windows can be written down with the aid of the formulas of Section A.6, but as they are rather complicated, we shall avoid doing so. The three choices eliminating constant trends, E_{0r}, E_{2r} and E_{4r}, all have integrated spectral windows approximating $1/n$. The dependence on r is in any case weak, and where present increases with r. For E_{0r} there is no dependence on r.

We are, of course, more concerned with the spectral windows associated with the modified V_r's rather than with those associated with the modifying E_{kr}. If we use E_{0r}, then the change in spectral window is the same for each lag. Consequently, only the $R_{30}(f)$ window is affected, the $R_{cr}(f)$ windows corresponding to the other V_r vanish. The situation is more complicated for the other cases, and no clear advantage over the use of E_{0r} appears.

The spectral window corresponding to $\bar{X}^2$ is, of course, (dif $nf/$dif $f)^2$ and is both always positive and well concentrated near zero. (If we replaced $\bar{X}$ by the average over a graded data window we could decrease the corresponding spectral window for f beyond the first side lobe of dif nf, but the concomitant broadening of the main lobe would make the result much less useful.) Clearly the use of E_{0r} will be quite satisfactory.

Comparing E_{1r}, E_{3r} and E_{5r} is not so simple. E_{3r}^* has the smaller integrated spectral window for r/n small, but the simplicity of calculation of E_{1r} will often outweigh this fact. (If economizing on the spectral window is important, we could use a windowless quadratic to eliminate the average effect of β.) Accordingly, the use of E_{1r} is recommended, unless it is simpler to subtract a fitted linear function from all X_t.

(There is no Section B.20.)

B.21 SAMPLE COMPUTING FORMULAS

Only the formulas for correction for prewhitening and for correction for the mean require discussion. The factor for $1 \leqq r \leqq m - 1$ is exactly the reciprocal of the prewhitening transfer function, calculated at the nominal frequency of the estimate. The factor for $r = m$, and the main portion of that for $r = 0$, differ only in the selection of the frequency at which the prewhitening transfer function is evaluated. Since, in each case, the nominal frequency is at one edge of the band of frequencies covered, the frequency of evaluation was displaced from the nominal frequency toward the center of the corresponding band. The choice of a point one-third of the way across the band was somewhat arbitrary.

Finally, there is the question of compensation for the correction for the mean. The raw estimate for $r = 0$ would naturally be thought of as corresponding to the interval from $-1/(4m)$ to $+1/(4m)$ cycles per observation, while the hanned estimate would cover from $-1/(2m)$ to $+1/(2m)$ cycles per observation. Since n degrees of freedom are associated with the entire interval from $-\frac{1}{2}$ to $+\frac{1}{2}$ cycle per observation, the hanned estimate for $r = 0$ is associated with n/m degrees of freedom. One of these has been eliminated by correction for the mean, as would also have been the case had we used E_{2r} or E_{4r}, so that we need to compensate for a reduction in the ratio

$$\frac{\dfrac{n}{m} - 1}{\dfrac{n}{m}} = \frac{n - m}{n}$$

whose reciprocal is the first compensating factor for $r = 0$. (Had we used E_{1r}, E_{3r}, or E_{5r}, we would have had to compensate for the loss of two degrees of freedom by a factor $n/(n - 2m)$.)

(There is no Section B.22.)

DETAILS FOR PLANNING

B.23 DURATION REQUIREMENT FORMULAS

We are now in a position to assemble and modify formulas from a number of sections as a basis for formulas expressing explicit requirements. In the process we shall have to give explicit definitions for certain concepts. The first of these is *resolution*. If we hann or hamm, we obtain estimates every $1/(2T_m)$ cps. Adjacent estimates have very considerably overlapping windows, and consequently the estimates have substantially related sampling fluctuations and refer to overlapping frequency regions. It would be a clear mistake to consider these estimates as completely resolved. When we come to next-adjacent estimates, however, the situation is quite different. The overlap is small, the covariance being about 5 per cent of either variance for a moderately flat spectrum. We shall consequently treat such pairs of estimates as completely resolved, and place

$$(\text{resolution in cps}) = 2\,\frac{1}{2T_m} = \frac{1}{T_m \text{ in seconds}}.$$

We can express the stability, so far most often expressed as the number of elementary frequency bands or equivalent degrees of freedom associated with each estimate, in terms of the spread in db of an interval containing, with prescribed probability, the ratio of true smoothed power to estimated smoothed power. Reference to Table II in Section 9 shows the inter-relations to be, approximately

$$k = 1 + \frac{250}{(80 \text{ per cent range in db})^2},$$

$$k = 1 + \frac{400}{(90 \text{ per cent range in db})^2},$$

$$k = 1 + \frac{625}{(96 \text{ per cent range in db})^2},$$

$$k = 1 + \frac{840}{(98 \text{ per cent range in db})^2}.$$

We shall write our combined formulas in terms of the 90 per cent range. Formulas for other per cent ranges are easily obtained by replacing 400 by the appropriate constant. It must be emphasized that when we use a 90 per cent range we only have 9 chances in 10 of finding *each* individual estimate correspondingly close to its average value and that if we have, say, 30 estimates, we are quite sure that at least one will be more discrepant than this.

We recall that we adopted (see end of Section 6 of Part I)

$$T'_n = (\text{total length of record}) - \frac{p}{3} T_m,$$

where p was the number of pieces, and, for design purposes.

$$k \approx \frac{2T'_n}{T_m}.$$

Hence,

$$(\text{duration in seconds}) = T_n = \left(\frac{T'_n}{T_m} + \frac{p}{3}\right) T_m$$

$$= \left(\frac{k}{2} + \frac{p}{3}\right) \frac{1}{(\text{resolution in cps})}$$

$$= \left(\frac{1}{2} + \frac{200}{(90 \text{ per cent range in db})^2} + \frac{(\text{pieces})}{3}\right) \Big/ (\text{resolution in cps}).$$

Writing (length of each piece)·(number of pieces) for (duration) and solving for the number of pieces yields

$$\text{(number of pieces)} = \frac{\dfrac{1}{2} + \dfrac{200}{(90 \text{ per cent range in db})^2}}{\text{(length of each piece)(resolution in cps)} - \frac{1}{3}}.$$

These relations are general, and apply equally to analog processed continuous records or digitally processed equi-spaced records, provided the spectrum is, or can be prewhitened to be, reasonably flat.

B.24 DIGITAL REQUIREMENT FORMULAS

If we are to use equi-spaced digital analysis, if we can provide frequency cutoff easily, and if we need to cover frequencies up to some $f_{\max}$, then we can probably take our folding frequency at about $\frac{3}{2} f_{\max}$. The necessary number of lags m then follow from

$$\Delta t = \frac{1}{3f_{\max}} = \frac{1}{2f_N},$$

$$m = \frac{T_m}{\Delta t} = 3T_m f_{\max}.$$

The necessary number of data points follows from

$$n = \frac{T_n}{\Delta t} = \left(T'_n + \frac{p}{3} T_m \right) 3f_{\max}$$

$$= 3T'_n f_{\max} + p T_m f_{\max}$$

$$= \left(3 \frac{T'_n}{T_m} + p \right) (T_m f_{\max}),$$

and the rough number of multiplications is

$$mn = 9T_m T'_n (f_{\max})^2 + 3p(T_m f_{\max})^2.$$

The quantity $T_m f_{\max}$ can be written as

$$\frac{f_{\max}}{1/T_m} = \frac{\text{maximum frequency}}{\text{resolution}} = \text{(number of resolved bands)},$$

so that the number of data points becomes

$$n = \left(3 \frac{T'_n}{T_m} + p \right) \text{(number of resolved bands)}$$

$$= (1.5k + p) \text{ (number of resolved bands)},$$

TABLE VII

Second differences of Brouwer's data and the add-and-subtract pilot estimation process as applied to them. (Block 2 only.)

Date	10F*	diff.	2nd diff. = X_q	δ	σ	$\delta\sigma$	$\sigma\sigma$	$\delta\sigma\sigma$	$\sigma\sigma\sigma$	DX_q
1853.5	56									
		0								
1854.5	56		−4							
		−4		4	−4					4
1855.5	52		0							
		−4				7	−1			−2
	48		−2							
		−6		7	3					7
	42		+5							
		−1						−2	−4	−9
	41		−4							
		−5		11	3					11
	36		+7							
		+2				−9	−3			−30
1860.5	38		−23							
		−21		40	−6					40
	17		+17							
		−4								−19
	13		−2							
		−6		−4	−8					−4
	7		−6							
		−12				18	2			+7
	−5		+1							
		−11		+8	10					+8
1865.5	−16		+9							
		−2						−23	−19	−20
	−18		−11							
		−13		+13	−9					+13
	−31		+2							
		−11				−3	−21			−1
	−42		+1							
		−10		−14	−12					−14
	−52		−13							
		−23								+15
1870.5	−75		+2							
		−21		+7	11					+7
	−96		+9							
		−12				−17	5			−20
	−108		−11							
		−23		+16	−6					+16
	−131		+5							
		−18						4	14	+4
	−149		+9							
		−9		−9	9					−9
1875.5	−158		0							
		−9				−9	9			+4
	−167		+4							
		−5		−8	0					−8
	−172		−4							
		−9								+8
	−181		+4							
		−5		−10	−2					−10
	−186		−6							
		−11				6	2			+15

TABLE VII, CONTINUED

Date	10F*	diff.	2nd diff. = X_q	δ	σ	$\delta\sigma$	$\sigma\sigma$	$\delta\sigma\sigma$	$\sigma\sigma\sigma$	LX_q
1880.5	−197		+9	−14	4					−14
		−2	−5							
	−199							1	5	−6
		−7	−11							
	−206			+33	11					+33
		−18	+22							
	−224					−19	3			−30
		+4	−8							
	−220			+8	−8					+8
		−4	0							
1885.5	−224									
		−4								
1886.5	−228									

* (Brouwer's notation) is the fluctuation in the earth's rotation. Here 10F is the fluctuation expressed in tenths of seconds of time.

and the rough number of multiplications becomes

$$nm = (4.5k + 3p)(\text{number of resolved bands})^2$$

$$= \left[\frac{9}{2} + \frac{1800}{(90 \text{ per cent range in db})^2} + 3(\text{pieces})\right]$$

$$\cdot (\text{number of resolved bands})^2.$$

These last formulas assume satisfactory shaping of frequency cutoff, and constant resolution up to the maximum frequency of interest. Particular situations may deviate from this in either direction.

(There are no Sections B.25, B.26, B.27.)

B.28 ANALYSIS OF EXAMPLE C

It is most desirable that an account of this sort include enough details of a numerical example to allow those readers who wish to do so, to follow through and check. This is impractical when even a few tenths (or even a few hundredths) of a million multiplications are involved. Example C, however, offers us an opportunity to present such details for one example, even if it is quite atypical.

We remarked in Section 28 that the add-and-subtract pilot estimation procedure of Section 18 *might* be applied to the second differences of Brouwer's values (themselves the differences ephemeris time *minus* mean solar time). During the 131 years from 1820 to 1950 astronomical techniques have improved, and observational errors seem, according to Brouwer's own analysis, to have somewhat decreased. In

TABLE VIII — SUMS OF SQUARED DIFFERENCES

	δ column	$\delta\sigma$ column	$\delta\sigma\sigma$ column	(*)	(†)
Block 1	8615	2083	321	−502	−971
Block 2	4130	1230	550	−9	−70
Block 3	3447	2591	273	1557	−244
Block 4	1553	304	516	−162	283
Noise	5/2	3/4	3/8	0	0
Flat	1	1	1	0.7	0.85

(*) Sum for $\delta\sigma$ column less 0.3 times sum for δ column.
(†) Sum for $\delta\sigma\sigma$ column less 0.15 times sum for δ column.

order to reflect this fact, and to provide some external estimate of error we shall study the second differences in 4 separate blocks, 4 separate time intervals of 32 consecutive years each. In view of the fact that Brouwer estimates the mean observational error to have decreased from 0.38 to 0.17 seconds of time over the period, it will clearly suffice to work in units of 0.1 second of time. Table VII shows, for the second time block, calendar dates, Brouwer's values, their first and second differences (the latter being the series we consider as the X_q) and the results of the add-and-subtract pilot estimation model. In this table, sums and differences are shown on lines half-way between the lines containing the entries from which they are formed. The last column shows the result of completing the δX_q column to a DX_q column. The resulting sums of squared differences are shown in Table VIII for each of the four blocks, together with comparative values.

These comparative values show the anticipated relative sizes of such sums of squared differences in case the spectral density were

(1) proportional to $(1 - \cos \pi f/f_N)^2 = (1 - u)^2$, the shape assumed for the "noise" component,

(2) flat, the shape assumed for the other component according to the second model,

where we have introduced u as an abbreviation for $\cos \pi f/f_N$. The successive columns have average sums of squared differences which are easily seen to be obtained by multiplication of the spectrum by (when we start with 32 X's):

$$16 (2 - 2 \cos \pi f/f_N) = 32 (1 - u),$$

$$8 (2 + 2 \cos \pi f/f_N) (2 - 2 \cos 2\pi f/f_N) = 64(1 + u - u^2 - u^3)$$

$$= 64 (1 + u) (1 - u^2),$$

$$4 (2 + 2u) (4u^2) (16 u^2 - 16 u^4) = 512(u^4 + u^5 - u^6 - u^7),$$

and integration.

When we recall that

$$\int_0^{f_N} u^k \, df$$

vanishes for odd k, and is equal to f_N, $f_N/2$, $3f_N/8$, $5f_N/16$ and $35f_N/128$ respectively, for $k = 0, 2, 4, 6$ and 8, we easily obtain the values given for comparison in Table VIII (after removing $32f_N$ as a common factor).

The last two columns of this table represent attempts to combine sums of squared differences so as to estimate the first component free of the "noise" component whose spectrum is proportional to $(1 - u)^2$. The attempts appear quite useless.

One reason for the lack of success is easy to find. There are only 4 values of $\delta\sigma\sigma$ to square and sum for each 32-year block. This means no more than 4 degrees of freedom, and, consequently very poor stability. We can partially correct this difficulty by going over to the modified add-and-subtract method in which all possible differences of a given sort are calculated.

Table IX presents the compact calculation for block 1, arranged as in Table VI. Here (σ) stands for first σ and then σ'. The mean squares for each column of differences lead to the results summarized in Table X for all four blocks. The difference estimates are now seen to be more stable and to increase somewhat with decreasing frequency, although not as much as would be expected even for a flat component.

The analysis is doing better, but is not yet satisfactory. One likely reason for this appears when we inquire what sort of spectral windows go with (*), (†) and (‡). The windows corresponding to $D(\sigma)$ and to (*) are shown in Fig. 24. The amount of negative area near $f/f_N = 1$ required to compensate for the rather large pickup of the "noise" component by the $D(\sigma)$ column is quite substantial, suggesting probable increased variability.

We can reduce our difficulties from such causes by using only slightly more complex processes. We can probably use the D column satisfactorily as an indication of the noise component.

We need to obtain two other composite measures of the spectrum, both of which avoid large values of f/f_N, one of which is concentrated near $f/f_N = 0$ and the other of which avoids $f/f_N = 0$. If we can obtain a smoothed and decimated sequence which avoids the upper part of the spectrum, then sums and differences of such values will have mean squares with the appropriate properties. The results of Section B.18 suggest trying $F_2 S_2 S_3$ as the operation generating the modified sequence, to which D (differencing) and S_2 are then to be applied.

TABLE IX

Compact calculation applied to second differences (in tenths of seconds of time) for block 1 (values in Brouwer's Table VIII(c) used where appropriate). (Entries arranged as in Table VI.)

X_q	D	(σ)	$D(\sigma)$	$(\sigma)^2$	$D(\sigma)^2$	$(\sigma)^3$	$D(\sigma)^3$	$(\sigma)^4$	$D(\sigma)^4$
-27		-6		-10		-10		-9	
	+48		2		10		11		13
+21		-4		0		+1		4	
	-27		4		-6		-5		
-6		0		-6		-4			
	+8		0		13		12		
+2		0		7		8		-3	
	-8		-1		-12				
-6		-1		-5					
	+12		-4		6				
+6		-5		1		-6		-4	
	+2		6		1		8		
+8		1		2		2			
	-16		5		4		1		
-8		6		6		3		. 5	
	+1		-10						
-7		-4							
	+13		3						
+6		-1		-4		-5		-7	
	0		3		3		3		
+6		2		-1		-2			
	-17		-3		-3		24		
-11		-1		-4		22		20	
	+16		23		6				
+5		22		2					
	-9		-42		-1				
-4		-20		1		-5		-2	
	+8		15		20		8		
+4		-5		21		3			
	-2		16		-46		-7		
+2		11		-25		-4		-1	
	+6								
+8									
	-20								
-12		15		11		10		26	
	+23		-19				16		6
∓11		-4		-1					
	-23		18		7		-23		9
-12		14		6		-7			
	+31				4				
+19		-15		10					
	-36		27				-4		
-17		12		6		5		21	
	+27				-19		11		
+10		-6		-13		16			
	-21		6				16		-18
-11		0		11		-2		14	
	+16		10						
+5		10							
	+12		-11						

TABLE IX, CONTINUED

X_q	D	(σ)	$D(\sigma)$	$(\sigma)^2$	$D(\sigma)^2$	$(\sigma)^3$	$D(\sigma)^3$	$(\sigma)^4$	$D(\sigma^4)$
+17		−1		10		7		10	
	−45		8		−13		−4		
−28		7		−3		3			
	+36		−14		−3		−20		
+8		−7		−6		−17		−14	
	+6		1		15				
+14		−6		9		−9			
	−33		−5						
−19		−11		0		−9		0	
	+28		33		−17		18		
+9		22		−17		9			
	−7		−32		29		−14		
+2		−10		12		−5		4	

TABLE X — MEAN SQUARES FOR DIFFERENCE COLUMNS AND SUITABLE
LINEAR COMBINATIONS THEREOF FOR THE ANALYSIS
ILLUSTRATED BY TABLE IX

Block	D	$D(\sigma)$	$D(\sigma)^2$	$D(\sigma)^3$	(*)	(†)	(‡)
1	485	283	234	176	−8	−57	−115
2	252	207	206	317	56	55	166
3	207	258	145	148	134	21	24
4	76	41	80	106	−5	34	60
Ave. of 2, 3 and 4....................					51	38	78
Noise	5/2	3/2	3/2	3/2	0	0	0
Flat	1	2	4	8	1.4	3.4	7.4

(*) Mean square for $D(\sigma)$ less 0.6 times that for D.
(†) Mean square for $D(\sigma)^2$ less 0.6 times that for D.
(‡) Mean square for $D(\sigma)^3$ less 0.6 times that for D.

The corresponding windows are, for the square of a single value of $S_2(F_2)S_2S_3X_q$,

$$4u^2(2 + 2u)(1 + 4u + 4u^2) = 8(u^2 + 5u^3 + 8u^4 + 4u^5),$$

and, for the square of a single value of $D(F_2)S_2S_3X_q$,

$$(4 - 4u^2)(2 + 2u)(1 + 4u + 4u^2) = 8(1 + 5u + 7u^2 - u^3 - 8u^4 - 4u^5),$$

to which should be compared that for the square of a single value of DX_q

$$(2 - 2u) = 2(1 - u).$$

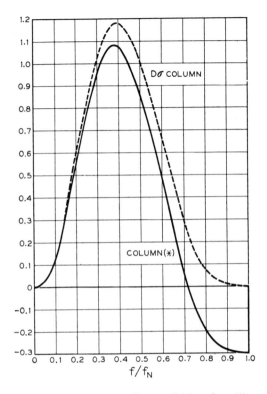

Fig. 24 — Windows corresponding to $D(\sigma)$ and to (*).

If now we form all values of $S_2(F_2)S_2S_3X_q$, $D(F_2)S_2S_3X_q$ and DX_q (Table XI illustrates the parallel calculation of F_2 and F_2' quantities), and form the corresponding mean squares, we obtain the results shown in Table XII.

If we compare the S-0.2δ column with the last column, we find a definite tendency for the last to be smaller. As indicated by the last two lines of the table, neither of these columns reflects, on the average, a "noise" component, while, on the average, both reflect a "flat" component equally. The natural conclusion is that the observed spectrum is more peaked toward $f/f_N = 0$ than would be expected from a mixture of "flat" and "noise" components. It might seem natural to conclude that the first model is to be preferred.

However, more careful examination shows that not only do the δ mean squares appear to decrease with improvements in astronomical technique, but (leaving aside block 1) the values in the S-0.2δ column are decreasing and those in the D-0.6δ column may well be doing the

TABLE XI

Calculation of $S_2(F_2)S_2S_3$ and $D(F_2)S_2S_3$ for the second differences of Brouwer's fluctuations in the earth's rotation. (Portion of block 2.)

Date	X_q	$S_2 X_q$	$F_2 S_2 S_2 X_q$	$DF_2 S_2 S_2 X_q$	$S_2 F_2 S_2 S_2 X_q$	$F_2' S_2 S_2 X_q$	$DF_2' S_2 S_2 X_q$	$S_2 F_2' S_2 S_2 X_q$
1854.5	−4							
1855.5	0	−6						
			−3					
	−2	+3						
				+10	+4	2		
	+5	−1						
			+7				−14	−10
	−4	+8						
				−26	−12	−12		
	+7	−20						
			−19				+5	−19
1860.5	−23	+1						
				+20	−18	−7		
	+17	−8						
			+1				+9	−5
	−2	+9						
				−4	−2	2		
	−6	−7						
			−3				+1	5
	+1	+4						
				+2	−4	3		
1865.5	+9	−1						
			−1				−11	−5
	−11	0						
				−17	−19	−8		
	+2	−8						
			−18				−12	−28
	+1	−10						
				+6	−30	−20		
	−13	−10						
			−12				18	−22
1870.5	+2	−2						
				15	−9	−2		
	+9	0						
			3				8	4
	−11	+3						
				14	20	6		
	+5	+3						
			17				21	33
	+9	+14						
				−4	30	27		
1875.5	0	+13						
			13				−23	31
	+4	0						
				−15	11	4		
	−4	+4						
			−2				−3	5
	+4	−6						
						1		
	−6	+7						
1880.5	9							

TABLE XII — MEAN SQUARES OF COLUMNS FOR SECOND
DIFFERENCES OF BROUWER'S DATA

For	S	D	δ	S-0.2δ	D-0.6δ	2.5 (D-0.6δ)
Block 1	83	228	484	−14	−62	−155
Block 2	290	178	252	240	27	68
Block 3	153	168	207	112	44	110
Block 4	62	47	75	47	2	5
Average (2, 3, 4)	168	131	178	133	24	61
Noise	1	3	5	0	0	0
Flat	28	12	2	27.6	10.8	27.6

S = mean square value of $S_2(F_2)S_2S_3X_q$.
D = mean square value of $D(F_2)S_2S_3X_q$.
δ = mean square value of DX_q .

same. The suggestion is — which, after we have seen another block or two or three (of 32 years each), we may be able to confirm or deny — that the lower frequency component, as well as the noise component, is decreasing as astronomical technique improves. If this be true, then most of the low-frequency power in blocks 2 and 3 may represent observational and reductional sources of variation rather than changes in the rotation of the earth.

While more refined analyses might show something more, 131 values are not a great number, especially since technique has changed during their measurement, and it is likely that we shall have to wait a while for a more definite answer to this question.

This example was included, not because of the peculiar importance of the irregularities in the rotation of the earth, but rather to illustrate certain general points, particularly these:

(1) While careful spectral analysis requires a very considerable amount of arithmetic, there are situations where simple analysis will yield useful results. (All the values in this section were obtained with pen or pencil and an occasional use of a slide rule. The use of simple differencing and summing techniques to produce moderately sharp windows has been studied by the Labroustes.[41, 42, 43] Although they felt themselves interested in line spectra, their methods, properly reinterpreted, are easily and directly applicable to continuous spectra.)

(2) The windows corresponding to the add-and-subtract pilot analysis are broad and the estimation far from exhaustive. (The procedure is intended for use in planning prewhitening, not as a tool for answering questions of even moderate difficulty.)

(3) Subdivision of the data before analysis is quite often helpful.

(If blocks 2, 3, and 4 had been analyzed as one unit, it would have been easy to jump to a conclusion which now seems dangerous.)

(4) Spectral analysis can lead to results unsuspected before the calculations were made. (The decrease in amplitude of the low-frequency component which Brouwer's data now suggests was not at all suspected until the values given in Table XII were pulled together.)

Index of Notations

In case a notation is not widely used, the Sections in which it is used are specified in parentheses:

a = an integer (A.6 and B.19),

a_0, a_1, a_2 = real constants (B-8.4),

a_{i0}, a_{ij} = assorted constants (13 and B.5),

A = various constants (27, B.9, B.15),

$A\left(f; \dfrac{1}{\Delta t}\right)$ = infinite Dirac comb made up of unit δ-functions spaced $1/\Delta t$ in frequency (see A.2 or Table IV for formula),

b = an integer (A.6 and B.19),

B = auxiliary quantity defined in B-8.5 (B.8),

$B(t)$, $B_i(t)$, $\tilde{B}(t)$ = a data window (graded) (10 and B.10),

c = an integer (A.6 and B.19),

c_0, c_1, $\cdots$, c_k = real constants (B.15),

C_{ij} = autocovariance corresponding to t_i and t_j (1),

$C(\tau)$ = autocovariance at lag τ,

$C_{00}(\tau)$ = autocovariance estimated from record of finite length,

$C'_{00}(\tau)$, $\tilde{C}_{00}(\tau)$, etc. = hypothetical or actual analogs of $C_{00}(\tau)$,

$C_i(\tau) = D_i(\tau)C_{00}(\tau)$ = modified autocovariance estimate (defined in 4),

C_r = sample autocovariance at lag $r\Delta\tau$ (usually = $r\Delta t$ in practice),

d = an integer (A.6, B.18 and B.19),

D = the operation of differencing adjoining values,

$D_i(\tau)$ = a prescribed even function of time shift, often a lag window (cp. 4, 5, B.4, B.5),

$D_{ei}(\tau)$ = lag window equivalent of $B_i(\tau)$ (see B-4.11 for representation),

e = 2.71828182845 $\cdots$,

E_{kr} = kth alternative correction at lag r (19, B.19),

f = frequency (in cycles per second where not otherwise specified), in equi-spaced discrete situations $f = r/(2m \cdot \Delta t) = (r/m)f_N$,

f_0 = a particular frequency (in Section B.18 the folding frequency of the cycle in question),

f_1 = a frequency, often the nominal frequency of a smoothed estimate,

f_N = Nyquist or folding frequency = $1/(2 \cdot \Delta t)$,

$f_N{}^*$ = effective Nyquist or folding frequency = $1/(2 \cdot \Delta\tau)$, (13),

F_j = the operation of retaining only every jth value $(j = 2, 3, \cdots)$ (B.19, B.28),

F_2' = the operation of retaining the alternate values dropped by F_2 (B.19, B.28),

g = an index running from $-c$ to $+d$ (A.6 and B.19),

$G(t)$ = a function of the time, Fourier transform of $S(f)$ (correspondence also required with subscripts 0, 1, 2),

$\tilde{G}(t)$ = the Fourier transform of a particular box-car function (B.12),

h = an integer, generally satisfying $\Delta\tau = h \cdot \Delta t$, in A.6 and B.19 an index running through indicated ranges,

$H_i(f; f_1) = Q_i(f + f_1) + Q_1(f - f_1)$ = power transfer function (from power at absolute frequency f to estimate at nominal frequency f_1),

$H_i(f - f_1)$ = special form of $H_i(f; f_1)$,

$H_Y(f; f_1)$ = power transfer function defined in B-10.2,

$i = \sqrt{-1}$,

Subscript i, values often 0, 1, 2, 3, 4, usually identifies quantities or functions associated with ith window pair,

j = an integer, often such that $3j - 2 = k$ or ℓ (B.19),

Subscript j, values often 0, 1, 2, 3, 4, alternative to subscript i,

k = usually number of equivalent degrees of freedom or number of elementary frequency bands; in Section 17, length of group averaged,

ℓ = an integer; in 18 an exponent of 2, in A.6 satisfying $n = 2\ell + 1$,

L = inductance (in example of 27),

m = integer, number of longest lag (longest lag = $m \cdot \Delta\tau$, usually = $m\Delta t$),

$M(t, \tau)$ = bilinear monomial in the X's
       = $X(t + \tau/2) \cdot X(t - \tau/2)$,

n = usually one less than the number of discrete data points, in A.6 and B.19 an integer = $2\ell + 1$,

n' = effective length of record $(n' \cdot \Delta t = T_n')$,

p = usually the number of pieces of record; also $= i\omega$ (A.1 only); a real number (usually integral, A.6),

p_0 = a constant power density (B.9),

$p_0, p_1, p_2, \cdots$ = ordinates in general example (9),

$P(f)$ = density of power spectrum (normalized so that variance = $\int_0^\infty 2P(f)\, df$), also with various subscripts,

$P_a(f)$ = aliased density of power spectrum (periodic in frequency, cp. 12),

$P_A(f)$ = principal part of aliased density of power spectrum (confined to $|f| \leqq 1/(2 \cdot \Delta\tau)$, cp. 12),

$P_Y(f_1)$ = estimate of spectral density based on filtered signal (B.10),

$P_{ei}(f)$ = spectral density estimated with the data window $B_i(t)$,

$P_{00}(f)$ = symbolic Fourier transform of $C_{00}(\tau)$,

$P_{0A}(f)$ = aliased estimate of smoothed $P(f)$,

$P_{i1}(f)$ = $H_i(f; f_1)P(f)$, roughly a filtered version of $P(f)$ (subscript i interchangeable with j, also 1 with 2),

$P_{iA}(f_1)$ = aliased estimate of smoothing of $P(f)$ by $H_i(f; f_1)$,

$P_{iA1}(f)$ = $H_i(f; f_1) \cdot P_A(f)$ = aliased filtered spectral density,

$P_{iAk}(f)$ = corrected aliased estimate of $P(f)$ smoothed by $Q_i(f)$,

$P_{\text{out}}(f; f_1)$ = $|Y(f; f_1)|^2 \cdot P(f)$ = spectral density of output,

$\tilde{P}(f)$ = a perfectly known power spectrum (B.12),

$\tilde{P}_A(f)$ = either the aliased spectrum corresponding to $\tilde{P}(f)$ (B.12) *or* the aliased spectrum of the $\tilde{X}_q$ series (15, 17),

$\tilde{P}_{i1}(f)$ = an auxiliary quantity defined in B-8.4,

q = index running as indicated,

Q_r = an auxiliary quantity (B.19),

$Q_0(f)$ = Fourier transform of centered box-car function of length $2T_m$ (See also $Q_i(f)$ for i),

$Q_i(f)$ = Fourier transform of $R_i(\tau)$ (see B.5 for i = 0, 1, 2, 3, 4 and B-8.7 for another choice),

$Q_0(f; \Delta\tau)$ = spectral window corresponding to (Fourier transform of) a discrete box-car function,

$\qquad = \Delta\tau \cdot \cos \frac{1}{2}(w \cdot \Delta\tau) \sin (mw \cdot \Delta\tau)$,

$Q_{ei}(f)$ = spectral window corresponding to (Fourier transform of) $D_{ei}(\tau)$,

$Q_{0A}(f)$ = aliased form of $Q_0(f)$ = $Q_0(f; \Delta\tau)$,

$Q_{iA}(f)$ = aliased form of $Q_i(f)$,

$\tilde{Q}_0(f), \tilde{Q}_i(f)$ = auxiliary quantities (B.8),

r = integer index almost always running from 0 or 1 to m or $m - 1$,

R = resistance (in example of Section 27),

$R_{ik}(f)$ = spectral window associated with $Q_i(f)$ and the sequence E_{k0}, $E_{k1}, \cdots, E_{km}$,

S_j = operation of summing by (overlapping) sets of j each (B.17, B.28),

$S(f)$ = function of frequency, Fourier transform of $G(t)$ (also with subscripts 1, 2, $\cdots$),

$\tilde{S}(f)$ = a particular box-car function (B.12),

t = time in the sense of epoch (also with various subscripts),

T = a time, usually positive,

T_m = half-length of box-car function of time, *or* greatest lag used or considered,

T_n = length of record,

T_n' = effective length of record, approximately $T_n - \dfrac{p}{3} T_m$,

u = general real variable,.

U_r = corrected estimate of smoothed power density (at nominal frequency $r/(2m \cdot \Delta t)$,

V_r = raw estimate of smoothed power density at nominal frequency $r/(2m \cdot \Delta t)$,

w = a chance quantity (B.6),

W = a frequency band width (B.9),

W_e = equivalent width, often of $P_{i1}(f)$ defined in Section 8,

W_{main} = width of main lobe,

W_{side} = width of side (unsplit) lobe,

$W(t)$ = impulse response of linear transmission system,

$W(t; f_1)$ = in B.10 the impulse response of filter with transfer function $Y(f; f_1)$,

x = usually a real variable, in B.6 a chance quantity (random variable),

X_q = qth value of discrete, equi-spaced time series,

$X(t)$ = value of time function,

$\bar{X}$ = in Section 1 an average value (along an infinite function or across an ensemble), in 19 ff. the mean of the observed X_q's,

$\overline{X^+}$, $\overline{X^-}$ = means of end thirds of observed values,

$\tilde{X}_q$, X_q^* = qth values of linearly transformed time series,

y = in Section B.6 a chance quantity (random variable),

$Y(f)$ = steady-state transfer function corresponding to linear transformation or linear transmission system,

z = in Section B.6 a chance quantity (random variable),

α = a real constant (15) *or* an indeterminate (B.8), *or* a certain fraction (B.18) *or* an unknown constant (B.19),

α_i = factor indicating extent of end effect losses (8, B.6) defined by B-6.9,

β = real constant (15) *or* an indeterminate (B.8) *or* a constant defined by B-8.8, *or* an unknown (slope) constant (B.19),

$\hat{\beta}$ = estimate of β(B.19),

γ = real constant (15),

$\Gamma(f)$ = power-variance spectral density,

$\Gamma_{\Delta t}(f)$ = power-variance spectral density in the equi-spaced discrete case,

δ = operation of forming alternate differences,

δ' = operation of forming alternate differences complementary to δ,

Δf = a change in frequency, sometimes the width $1/(2T_n')$ of an elementary frequency band,

Δt = time interval, usually that between data values,

$\Delta \tau$ = time interval, usually that between lags used ($= h \cdot \Delta t$),

ϵ = a small real number (A.2),

ξ = real, time-like variable of integration (A.3),

λ = real, time-like variable of integration (A.3) *or* a real constant (15),

μ = real constant (15),

ν = real constant (15),

ω = angular frequency (in radians per second unless otherwise specified) always = $2\pi f$ [with any sub- or superscripts],

φ = $2\pi f T_m$, a normalized frequency (B.8),

$\Phi(f, \lambda)$ = an auxiliary function (B.6) (defined in B-6.3),

ψ = a phase angle (A.6),

χ_k^2 = a quantity distributed as chi-square on k degrees of freedom (9),

σ, σ' = complementary operations of summing by adjacent parts and then omitting every alternate sum,

τ = time difference or lag,

$*$ = sign of convolution,

superscript $*$ = sign of complex conjugate (A.3),

$\nabla(t; \Delta t)$ = infinite Dirac comb approximating the constant unity [formula in A.2 or Table IV],

$\nabla_m(t; \Delta t)$ = finite Dirac comb approximating a unit-height, centered, box-car function [formula in A.2 or Table IV].

Glossary of Terms

Add-and-subtract method

A method of roughly estimating spectra based on successive additions by non-overlapping two's followed by a differencing. (18, B.18 and B.28.)

Alias

In equally spaced data, two frequencies are aliases of one another if sinusoids of the corresponding frequencies cannot be distinguished by their equally spaced values (this occurs when $f_1 = 2kf_N \pm f_2$ for integer k); the *principal aliases* lie in the interval $-f_N \leqq f \leqq f_N$. (See also 12.) (Also *aliased, aliasing*, etc.)

Aliased spectrum

See Spectrum, aliased.

Analysis, pilot

Any of a number of methods of obtaining a rough spectrum, including the add-and-subtract method (18, etc.) the cascade method (B.18), the complete add-and-subtract method (B.18).

Autocorrelation function

The normalized autocovariance function (normalized so that its value for lag zero is unity).

Autocovariance function

The covariance between $X(t)$ and $X(t + \tau)$ as a function of the lag τ. If averages of $X(t)$ and $X(t + \tau)$ are zero, it is equal to the average value of $X(t) \cdot X(t + \tau)$. It can be defined for a whole ensemble, a whole function stretching from $-\infty$ to $+\infty$, or for a finite piece of a function; in the latter case it is called the *apparent autocovariance function* (see 4). Certain related functions are called modified apparent autocovariance functions (also see 4).

Autoregressive series

A time series generated from another time series as the solution of a linear difference equation. (Usually where previous values of output series enter into determination of current value.)

Average

The arithmetic mean, usually over an ensemble, a population, or some reasonable facsimile thereof.

Band-limited function

Strictly, a function whose Fourier transform vanishes outside some finite interval (and hence is an entire function of exponential type); practically, a function whose Fourier transform is very small outside some finite interval.

Box-car function

A function zero except over a finite interval, in the interior of which it takes a constant value (often $+1$).

Cardinal theorem (of interpolation theory)

A precise statement of the conditions under which values given at a doubly infinite set of equally spaced points can be interpolated (with the aid of the function $(\sin (x - x_i))/(x - x_i)$ to yield a continuous band-limited function. (See B.1.)

Cascade process (of spectral estimation)

A process of spectral estimation in which a single step is repeated again and again, each step yielding both certain estimates and a condensed set of data (ready for input to the next step). (See B.18.)

Chi-square

A quantity distributed (strictly exactly, but practically approximately) as $x_1^2 + x_2^2 + \cdots + x_k^2$ where $x_1, x_2, \cdots, x_k$ are independent and Gaussian, and have average zero and variance unity.

Continuous power spectrum

A power spectrum representable by the indefinite integral of a suitable (spectral density) function. (All power spectra of physical systems are continuous.)

Convolution

The operation on one side of a Fourier transformation corresponding to multiplication on the other side. (See A.3 for detailed discussion.)

Cosine transform

A series (see 13) or integral (see 2) transform in which a cosine of the product of the variables is the kernel.

Covariance

A measure of (linear) common variation between two quantities, equal to the average product of deviations from averages. (See 1.)

Cross-spectrum

The expression of the mutual frequency properties of two series analogous to the spectrum of a single series. (Because mutual relations at a single frequency can be in phase, in quadrature, or in any mixture of these, either a single complex-valued cross-spectrum or a pair of real-valued *cross-spectra* are required.) (Also *cross-spectral*.)

Data

As specifically used in this paper, values given at equally spaced intervals of time (often called time series).

Data window

A time function which vanishes outside a given interval and which is regarded as multiplying data or signals defined for a more extended period. (Data windows are usually smooth (graded) to improve the quality of later frequency analysis.)

Degrees of freedom

As applied to chi-square distributions arising from quadratic forms in Gaussian (normal) variables, the number of linearly independent squared terms of equal size into which the form can be divided. In general, a measure of stability equal to twice the square of the average divided by the variance.

Delta-component

A finite contribution to the spectrum at one frequency (B.10 only).

Diffraction function

$$\mathrm{dif}\ x \equiv \frac{\sin \pi x}{\pi x}.$$

Dirac comb

An array of equally spaced Dirac functions, usually most of which are of equal height.

Dirac function

The limit of functions of unit integral concentrated in smaller and smaller intervals near zero. (See A.2 for fuller discussion.)

Distortion

Failure of output to match input. (Often specified as to kind of failure as *linear, amplitude, phase, non-linear*, etc., cp. A.3.)

Effective record length

Actual length of record available reduced to allow for end effects. (See 6.)

Elementary frequency band

An interval of frequency conveniently thought of as containing "a single degree of freedom", equal to the reciprocal of twice the duration of observation or record. (Since both sines and cosines may occur, it requires *two* elementary frequency bands to contain "an independently observable frequency.")

Ensemble

A family of functions (here functions of either continuous or equi-spaced time) with probabilities assigned to relevant sub-families.

Equivalent number (of degrees of freedom)

See second sentence under *degrees of freedom*.

Equivalent width

The extent of a function regarded as a window as expressed by the ratio of the square of its integral to the integral of its square. (See 8.)

Filtered spectrum

Spectrum of the output from any process which can be regarded as a filter.

Folding frequency

The lowest frequency which "is its own alias", that is, is the limit of both a sequence of frequencies and of the sequence of their aliases, given by the reciprocal of twice the time-spacing between values, also called *Nyquist* frequency.

Fourier transform

Operations making functions out of functions by integration against a kernel of the form exponential function of $\sqrt{-1}$ times frequency times time. Often, including here, defined differently for transforming time functions into frequency functions than for transforming frequency functions into time functions. (See A.1 for details.)

Frequency

A measure of rate of repetition; unless otherwise specified, the number of cycles per second. The *angular frequency* is measured in radians per second, and is, consequently, larger by a factor of 2π.

Frequency window

The Fourier transform of a data window.

Gaussian

A single quantity, or a finite number of quantities distributed according to a probability density representable as e to the power minus a quadratic form. (Also called *normal*, *Maxwellian*, etc.) Also, a function or ensemble, distributed in such a way that all finite sections are Gaussian. (See 1.)

Hamming

The operation of smoothing with weights 0.23, 0.54 and 0.23. (After R. W. Hamming.)

Hanning

The operation of smoothing with weights 0.25, 0.50 and 0.25. (After Julius von Hann.)

Hyperdirective antenna

An antenna or antenna system so energized as to have a more compact directional pattern than naturally corresponds to its extent (as measured in wavelengths).

Impulse response

The time function describing a linear system in terms of the output resulting from an input described by a Dirac function.

Independence (statistical, of estimates)

In general, two quantities are statistically independent if they possess a joint distribution such that (incomplete or complete) knowledge of one does not alter the distribution of the other. Estimates are statistically independent if this property holds for each fixed true situation.

Independent phases

An ensemble has independent phases when it can be approximated by ensembles consisting of finite sums of (phased) cosines (of fixed frequencies) whose phases are mutually independent. Continuous spectrum and independent phases imply Gaussian character. Every Gaussian ensemble has independent phases.

Intermodulation distortion

Non-linear distortion, especially as recognized in the output of a system when two or more frequencies enter the input simultaneously.

Joint probability distribution

Expression of the probability of simultaneous occurrence of values of two or more quantities.

Lag

A difference in time (epoch) of two events or values considered together.

Lag window

A function of lag, vanishing outside a finite interval, and either multiplying or regarded as multiplying the quantities of a family of quantities with differing lags.

Lagged product

The product of two values corresponding to different times. (In a *mean lagged product* the lags are usually all the same.)

Lead

The negative of lag.

Line (in a power spectrum)

Theoretically, and as used in this paper, a finite contribution associated with a single frequency. Physically, not used here, a finite contribution associated with a very narrow spectral region.

Lobe

A bulge, positive or negative, especially in a spectral window. (In most spectral windows, a large central *main lobe* is surrounded on both sides by smaller *side lobes*.)

Mean lagged product

The (arithmetic) mean of products of equally lagged quantities.

Moving linear combination

A transformation expressing the values of an output time series as linear combinations of values of the input series in specified relations of lag (or lead).

Negative frequencies

When sines and cosines are jointly represented by two imaginary exponentials, one has a positive frequency and the other a negative frequency. (Not specifiable for a single time function in real terms.)

Network (linear)

In this account, an otherwise unspecified physical device which converts an input function (of continuous time) *linearly* into an output function (of continuous time).

Noise

In general, an undesired time-function, or component of a function.

Non-normality

Failure to follow a normal or Gaussian distribution.

Normality

The property of following a normal or Gaussian distribution.

Nyquist frequency

The lowest frequency coinciding with one of its own aliases, the reciprocal of twice the time interval between values (same as *folding frequency*).

Octave

An interval of frequencies, the highest of which is double the lowest.

Pilot (analysis or estimation)

A process yielding rough estimates of spectral density intended mainly as a basis for planning more complete and precise analyses.

Population

A collection of objects (in particular, of numbers or of functions), with probabilities attached to relevant subcollections.

Power transfer function

The function expressing the ratio of output power near a given frequency to the input power near that frequency.

Power-variance spectrum

A function of frequency, in terms of which the variances and covariances of a family of spectral estimates can be expressed in standard form. (See 6 and 14 for details in the continuous and equi-spaced cases, respectively.)

Preemphasis

Emphasis of certain frequencies (in comparison with others), before processing, as an aid to the quality of result.

Prewhitening

Preemphasis designed to make the spectral density more nearly constant (the spectrum more nearly flat).

Principal alias

An alias falling between zero and plus or minus the folding or Nyquist frequency.

Process (random or stochastic)

An ensemble of functions. (Often composed of functions of time regarded as unfolding or developing.)

Protection ratio

The ratio of transmission at a desired frequency to the transmission at an undesired alias of that frequency.

Recording

Is *spaced* when originally taken at equal intervals, *mixed* when taken continuously and processed at equal intervals, *continuous* when taken and processed on a continuous basis.

Resolution

A measure of the concentration of a spectral estimate expressed in frequency units, here taken (for the important cases) as equal to the half-width, $1/T_m$, of the major lobe. (See B.23.)

Resolved bands (number of)

The ratio of the Nyquist or folding frequency to the resolution.

Sampling theorem (of information theory)

Nyquist's result that equi-spaced data, with two or more points per cycle of highest frequency, allows reconstruction of band-limited functions. (See *Cardinal theorem*.)

Serial correlation coefficients

Ratios of the autocovariances to the variance of a process, ensemble, etc.

Signal

A time function desired as (potentially) carrying intelligence.

"Signal"

A function of continuous time, which may be either a signal, a noise, or a combination of both. (Contrasted with *data*, a function of discrete time.)

Single function approach

A mode of representing certain ensembles by the translations of a single time function (in single function terms).

Smoothed function

The result of weighted averaging of nearby values of the original function.

Smoothing

In the narrow sense, forming (continuous or discrete) moving linear combinations with unit total weight.

Smoothing and decimation procedure

A procedure which may be regarded as the formation of discrete moving linear combinations, followed by the omission of all but every kth such. (See 17 and B.17.)

Spectrum (also power spectrum)

An expression of the second moments of an ensemble, process, etc. (i) in terms of frequencies, (ii) in such a form as to diagonalize the effects on second moments of time-invariant linear transformations applied to the ensemble or process. (adjective: *spectral*).

Spectrum, aliased

For equally spaced data, the *principal part* of the *aliased spectrum* expresses contributions to the variance in terms of frequencies between zero and the Nyquist or folding frequency, all contributions from frequencies having the same principal alias and sign having been combined by addition. (The *aliased spectrum* repeats the principal part periodically with period $2f_N$. See 14.)

Spectral density

A value of a function (or the entire function) whose integral over any frequency interval represents the contribution to the variance from that frequency interval.

Spectral density estimates

Estimates of spectral density, termed *raw* when obtained from equispaced mean lagged products by cosine series transformation, *refined* when hanned or hammed from raw estimates or obtained by an equivalent process. (See B.13.)

Spectral window

A function of frequency expressing the contribution of the spectral density at each frequency to the average value of an estimate of (smoothed) spectral density.

Stationary (ensemble or random process)

An ensemble of time functions (or random process) is stationary if any translation of the time origin leaves its statistical properties unaffected.

Superposition theorem

A statement that the output of a linear device is the convolution of its input with its impulse response. (See A.3.)

Temporally homogeneous

Sometimes used in place of *stationary*, especially when speaking of stochastic processes.

Transfer function

The *transfer function* of a network or other linear device is a complex-valued function expressing the amplitude and phase changes suffered by cosinusoidal inputs in becoming outputs. (See A.3.) The square of the absolute value of the transfer function is the *power transfer function*, which expresses the factors by which spectral densities are changed as inputs become outputs. (See 4.)

Transmission

The coefficient with which power at a given frequency contributes to power at the (new) principal alias as a result of the application of a smoothing and decimation procedure.

Transversal filtering

Time domain filtering by forming linear combinations of lagged values, use of moving linear combinations for filtering. (See Kallmann[32] for the origin of this term.)

Trend

A systematic, smooth component of a time function (time series), as, for example, a linear function of time (a *linear trend*).

True

Often used to refer to average values over the ensemble, as contrasted with mean values over the observations.

Universe

A collection of objects (numbers, functions, etc.) with probabilities attached to relevant subcollections.

Variance

A quadratic measure of variability, the average squared deviation from the average.

White noise

An ensemble whose spectral density is sensibly constant from zero frequency through the frequencies of interest (in equi-spaced situations, up to the folding or Nyquist frequency). (The values of equi-spaced white noise at different times are independent.)

Window

A function expressing, as a multiplicative factor, the tendency or possibility of the various values of some function to enter into some calculation or contribute to the average value of some quantity. (See *data, lag, spectral*, etc. for specific instances.)

Windowless quadratic

A quadratic expression is windowless if its average value vanishes for every stationary ensemble of finite variance (See B.19).

Window pair

Two windows related by a Fourier transformation, as lag and spectral windows or data and frequency windows. (See A.4 and 4.)

Zero-frequency waves (cosine and sine)

The limiting forms of very-low-frequency cosinusoids, namely constants and linear trends. (See 19.)

ACKNOWLEDGEMENT

Many persons have contributed by discussion, questioning and suggestion to the development of these ideas and techniques. An attempt to list all those who have helped during the eight years or so of growth and development could not succeed. Thus, we must express our thanks, generally, to many colleagues interested in transmission theory, statistics or computation, here in Bell Telephone Laboratories, and at Princeton and other universities, and to many scientists and engineers who have used the techniques to solve their own problems, and by real examples have shown us what had to be considered. Special thanks must go to Miss Dorothy T. Angell for patient and careful computations during numerical exploration, and to R. W. Hamming for continuing active discussion, especially of computational and expository problems. A careful reading of the final draft by S. O. Rice led to the elimination of many slips and ambiguities (of course those which remain are the authors' sole responsibility).

REFERENCES (PARTS I AND II)

1. N. R. Goodman, *On the Joint Estimation of the Spectra, Cospectrum and Quadra-ture Spectrum of a Two-Dimensional Stationary Gaussian Process*, Scientific Paper No. 10, Engineering Statistics Laboratory, New York University, 1957. (Ph.D. Thesis, Princeton University).
2. G. I. Taylor, Statistical Theory of Turbulence, Proc. Roy. Soc. (London), **A151,** pp. 421–478, 1935.
3. S. O. Rice, Mathematical Analysis of Random Noise, B.S.T.J., **23,** pp. 282–332, July, 1944; **24,** pp. 46–156, Jan. 1956. (This article also appears in *Selected Papers on Noise and Stochastic Processes*, edited by N. Wax, Dover Publ., New York, 1954, pp. 133–294.)
4. J. W. Tukey, The Sampling Theory of Power Spectrum Estimates, in *Symposium on Applications of Autocorrelation Analysis to Physical Problems*, Woods Hole, June 13, 1949. NAVEXOS-P-735, Office of Naval Research.
5. M. S. Bartlett, Periodogram Analysis and Continuous Spectra, Biometrika, **37,** pp. 1–16, 1950.
6. W. Marks and W. J. Pierson, The Power Spectrum Analysis of Ocean-Wave Records, Trans. Am. Geophysical Union, **33,** pp. 834–844, 1952.
7. H. Press and J. C. Houbolt, Some Applications of Generalized Harmonic Analysis to Gust Loads on Airplanes, J. Aero. Sciences, **22,** pp. 17–26, 1955.
8. H. A. Panofsky, Meteorological Applications of Power Spectrum Analysis, Bull. Am. Meteorological Soc., **36,** pp. 163–166, 1955.
9. G. P. Wadsworth, E. A. Robinson, J. G. Bryan, and P. M. Hurley, Detection of Reflections on Seismic Records by Linear Operators, Geophysics, **18,** pp. 539–586, 1953. (Also see later reports, Geophysical Analysis Group, M.I.T.)
10. M. S. Bartlett and J. Mehdi, On the Efficiency of Procedures for Smoothing Periodograms from Time Series with Continuous Spectra, Biometrika, **42,** pp. 143–150, 1955.
11. M. S. Bartlett, *An Introduction to Stochastic Processes*, Cambridge Univ. Press, Cambridge, England, 1955.
12. U. Grenander and M. Rosenblatt, *Statistical Analysis of Stationary Time Series*, Wiley, New York, 1957.
13. J. W. Tukey and C. P. Winsor, *Note on Some Chi-Square Normalizations*, Memorandum Report 29, Statistical Research Group, Princeton, 1949.
14. G. Groves, Numerical Filters for Discrimination Against Tidal Periodicities, Trans. Am. Geophysical Union, **36,** pp. 1073–1084, 1955.
15. H. R. Seiwell, The Principles of Time Series Analysis Applied to Ocean Wave Data, Proc. Nat. Acad. Sciences, **35,** pp. 518–528, Sept., 1949.
16. H. R. Seiwell and G. P. Wadsworth, A New Development in Ocean Wave Research, Science, **109,** pp. 271–274, 1949.
17. D. Brouwer, A Study of Changes in the Rate of Rotation of the Earth, Astronomical J., **57,** pp. 126–146, 1952.
18. Brouwer, pp. 135, 138.
19. G. A. Campbell and R. M. Foster, Fourier Integrals for Practical Applications, B.S.T.J., **7,** pp. 639–707, 1928; also, Van Nostrand, New York, 1942.
20. E. A. Guillemin, *Introductory Circuit Theory*, Wiley, New York, 1953,p. 257.
21. L. Schwartz, *Theorie des Distributions*, Vols. I and II, Hermann et Cie, Paris, 1950. (See also Halperin; Friedman; and Clavier, noted in Bibliography.)
22. E. A. Guillemin, *The Mathematics of Circuit Analysis*, Wiley, 1949, pp. 485–501.
23. H. Nyquist, Certain Topics in Telegraph Transmission Theory, Trans. A.I.E.E., pp. 617–644, April, 1928.
24. J. M. Whittaker, *Interpolatory Function Theory*, Cambridge Univ. Press, Cambridge, England, 1935, Chapter IV.
25. W. R. Bennett, Methods of Solving Noise Problems, Proc. I.R.E., **44,** 1956.
26. R. W. Hamming and J. W. Tukey, *Measuring Noise Color*, unpublished memorandum.
27. C. L. Dolph, A Current Distribution for Broadside Arrays Which Optimizes the Relationship between Beam Width and Side-Lobe Level, Proc. I.R.E., **34,** pp. 335–348, 1946.
28. P. Jaquinot, Quelques recherches sur les raies faible dans les spectres optiques, Proc. Phys. Soc. (London), **63B,** pp. 969–979, 1950.
29. P. Boughon, B. Dossier, and P. Jaquinot, Determination des fonctions pour

l'amelioration des figures de diffraction dans le spectroscope, C. R. Acad. Sci. Paris, **223**, pp. 661–663, 1946.

30. L. Isserlis, On a Formula for the Product-Moment Coefficient of Any Order of a Normal Frequency Distribution in Any Number of Variables, Biometrika, **12**, pp. 134–139, 1918.
31. H. Hotelling, Relations Between Two Sets of Variates, Biometrika, **28**, pp. 321–377, 1936.
32. H. E. Kallmann, Transversal Filters, Proc. I.R.E., **28**, pp. 302–310, 1940.
33. H. Wold, *A Study in the Analysis of Stationary Time Series*, 2nd Edition, Almquist and Wiksell, Stockholm, 1954.
34. Daniel B. DeLury, *Values and Integrals of the Orthogonal Polynomials up to $n = 26$*, Toronto, University Press, 1950.
35. Ronald A. Fisher and Frank Yates, *Statistical Tables for Biological, Agricultural and Medical Research*, Edinburgh, Oliver and Boyd, 4th Ed. 1953, (especially Table XXII).
36. William E. Milne, *Numerical Calculus*, Princeton University Press, 1949.
37. Cornelius Lanczos *Tables of Chebyshëv Polynomials $S_n(x)$ and $C_n(x)$*, Nat'l. Bur. Stds. App'd. Math. Series 9, Washington, Gov't. Print. Off., 1952, (especially pp. xiv to xvi).
38. Cornelius Lanczos, *Trigonometric Interpolation of Empirical and Analytical Functions*, J. Math. Phys., **17**, pp. 123–199, 1938.
39. Robert C. Minnick, *Tshebycheff Approximations for Power Series*, J. Assoc. Comp. Machinery, **4**, pp. 487–504, 1957.
40. J. C. P. Miller, *Two Numerical Applications of Chebyshëv Polynomials*, Proc. Roy. Soc. Edinb., **62**, pp. 204–210, 1946.
41. H. Labrouste and Y. Labrouste, Harmonic Analysis by Means of Linear Combinations of Ordinates, J. Terr. Magnetism, **41**, 1936.
42. H. Labrouste and Y. Labrouste, Analyse des graphiques resultant de la superposition de sinusoides, Mem. Acad. Sci. Inst. France, (2), **64**, No. 5, 1941. (See also Ann. de l'Inst. de Phys. du Globe de Univ. de Paris, **7**, pp. 190–207; **9**, pp. 99–101; **11**, pp. 93–101; **14**, pp. 77–91, for much of this material.)
43. H. Labrouste and Y. Labrouste, *Tables Numeriques precedes d'un expose de la methode d'analyse par combinaison d'ordonnees*, Presses Univ., Paris, 1943.

BIBLIOGRAPHY

In addition to the articles cited in the text, the reader's attention is called to the additional titles below.

P. Boughon, B. Dossier, and P. Jaquinot, Apodisation des raies spectrales au moyen d'ecrans absorbents, J. Recherches C.N.R.S., No. **11**, pp. 49–69, 1950.

P. A. Clavier, Some Applications of the Laurent Schwarz Distribution Theory to Network Problems, *Proceedings of the Symposium on Modern Network Synthesis, II*, Polytechnic Institute of Brooklyn, 1955.

L. Couffignal, La Methode de H. Labrouste pour la recherche des periodes, Ciel et Terre, **66**, pp. 78–86, 1950.

B. Friedman, *Principles and Techniques of Applied Mathematics*, Wiley, New York, 1956, Chapter 3.

I. Halperin, *Introduction to the Theory of Distributions*, University of Toronto Press, Toronto, 1952.

B. Mazelsky, Extension of the Power Spectrum Methods of Generalized Harmonic Analysis to Determine Non-Gaussian Probability Functions of Random Input Disturbances and Output Responses of Linear Systems, J. Aeronautical Sciences, **21**, pp. 145–153, March, 1954.

H. A. Panofsky and I. van der Horen, Spectra and Cross-Spectra of Velocity Components in the Mesometeorological Range, Quart. J. Roy. Meteorol. Soc., **81**, pp. 603–605, 1955.

H. A. Panofsky and R. A. McCormack, The Vertical Momentum Flux at Brookhaven at 109 Meters, Geophys. Res. Papers (USAF, Cambridge Research Center), No. **19**, pp. 219–230, 1952.

H. Press and J. W. Tukey, Power Spectral Methods of Analysis and Their Application to Problems in Airplane Dynamics, *Flight Test Manual, NATO, Advisory Group for Aeronautical Research and Development*, IV-C, pp. 1–41, June, 1956. (Reprinted as Bell System Monograph No. 2606.)

Index

C

D

CATALOGUE OF DOVER BOOKS

ENGINEERING AND TECHNOLOGY

General and mathematical

ENGINEERING MATHEMATICS, Kenneth S. Miller. A text for graduate students of engineering to strengthen their mathematical background in differential equations, etc. Mathematical steps very explicitly indicated. Contents: Determinants and Matrices, Integrals, Linear Differential Equations, Fourier Series and Integrals, Laplace Transform, Network Theory, Random Function . . . all vital requisites for advanced modern engineering studies. Unabridged republication. Appendices: Borel Sets; Riemann-Stieltjes Integral; Fourier Series and Integrals. Index. References at Chapter Ends. xii + 417pp. 6 x 8½.   S1121 Paperbound **$2.00**

MATHEMATICAL ENGINEERING ANALYSIS, Rufus Oldenburger. A book designed to assist the research engineer and scientist in making the transition from physical engineering situations to the corresponding mathematics. Scores of common practical situations found in all major fields of physics are supplied with their correct mathematical formulations—applications to automobile springs and shock absorbers, clocks, throttle torque of diesel engines, resistance networks, capacitors, transmission lines, microphones, neon tubes, gasoline engines, refrigeration cycles, etc. Each section reviews basic principles of underlying various fields: mechanics of rigid bodies, electricity and magnetism, heat, elasticity, fluid mechanics, and aerodynamics. Comprehensive and eminently useful. Index. 169 problems, answers. 200 photos and diagrams. xiv + 426pp. 5⅜ x 8½.   S919 Paperbound **$2.00**

MATHEMATICS OF MODERN ENGINEERING, E. G. Keller and R. E. Doherty. Written for the Advanced Course in Engineering of the General Electric Corporation, deals with the engineering use of determinants, tensors, the Heaviside operational calculus, dyadics, the calculus of variations, etc. Presents underlying principles fully, but purpose is to teach engineers to deal with modern engineering problems, and emphasis is on the perennial engineering attack of set-up and solve. Indexes. Over 185 figures and tables. Hundreds of exercises, problems, and worked-out examples. References. Two volume set. Total of xxxiii + 623pp. 5⅜ x 8.
S734 Vol I Paperbound **$1.85**
S735 Vol II Paperbound **$1.85**
The set **$3.70**

MATHEMATICAL METHODS FOR SCIENTISTS AND ENGINEERS, L. P. Smith. For scientists and engineers, as well as advanced math students. Full investigation of methods and practical description of conditions under which each should be used. Elements of real functions, differential and integral calculus, space geometry, theory of residues, vector and tensor analysis, series of Bessel functions, etc. Each method illustrated by completely-worked-out examples, mostly from scientific literature. 368 graded unsolved problems. 100 diagrams. x + 453pp. 5⅝ x 8⅜.   S220 Paperbound **$2.00**

THEORY OF FUNCTIONS AS APPLIED TO ENGINEERING PROBLEMS, edited by R. Rothe, F. Ollendorff, and K. Pohlhausen. A series of lectures given at the Berlin Institute of Technology that shows the specific applications of function theory in electrical and allied fields of engineering. Six lectures provide the elements of function theory in a simple and practical form, covering complex quantities and variables, integration in the complex plane, residue theorems, etc. Then 5 lectures show the exact uses of this powerful mathematical tool, with full discussions of problem methods. Index. Bibliography. 108 figures. x + 189pp. 5⅜ x 8.
S733 Paperbound **$1.35**

Aerodynamics and hydrodynamics

AIRPLANE STRUCTURAL ANALYSIS AND DESIGN, E. E. Sechler and L. G. Dunn. Systematic authoritative book which summarizes a large amount of theoretical and experimental work on structural analysis and design. Strong on classical subsonic material still basic to much aeronautic design . . . remains a highly useful source of information. Covers such areas as layout of the airplane, applied and design loads, stress-strain relationships for stable structures, truss and frame analysis, the problem of instability, the ultimate strength of stiffened flat sheet, analysis of cylindrical structures, wings and control surfaces, fuselage analysis, engine mounts, landing gears, etc. Originally published as part of the CALCIT Aeronautical Series. 256 Illustrations. 47 study problems. Indexes. xi + 420pp. 5⅜ x 8½.
S1043 Paperbound **$2.25**

FUNDAMENTALS OF HYDRO- AND AEROMECHANICS, L. Prandtl and O. G. Tietjens. The well-known standard work based upon Prandtl's lectures at Goettingen. Wherever possible hydrodynamics theory is referred to practical considerations in hydraulics, with the view of unifying theory and experience. Presentation is extremely clear and though primarily physical, mathematical proofs are rigorous and use vector analysis to a considerable extent. An Enginering Society Monograph, 1934. 186 figures. Index. xvi + 270pp. 5⅜ x 8.
S374 Paperbound **$1.85**

FLUID MECHANICS FOR HYDRAULIC ENGINEERS, H. Rouse. Standard work that gives a coherent picture of fluid mechanics from the point of view of the hydraulic engineer. Based on courses given to civil and mechanical engineering students at Columbia and the California Institute of Technology, this work covers every basic principle, method, equation, or theory of interest to the hydraulic engineer. Much of the material, diagrams, charts, etc., in this self-contained text are not duplicated elsewhere. Covers irrotational motion, conformal mapping, problems in laminar motion, fluid turbulence, flow around immersed bodies, transportation of sediment, general charcteristics of wave phenomena, gravity waves in open channels, etc. Index. Appendix of physical properties of common fluids. Frontispiece + 245 figures and photographs. xvi + 422pp. 5⅜ x 8. S729 Paperbound **$2.25**

WATERHAMMER ANALYSIS, John Parmakian. Valuable exposition of the graphical method of solving waterhammer problems by Assistant Chief Designing Engineer, U.S. Bureau of Reclamation. Discussions of rigid and elastic water column theory, velocity of waterhammer waves, theory of graphical waterhammer analysis for gate operation, closings, openings, rapid and slow movements, etc., waterhammer in pump discharge caused by power failure, waterhammer analysis for compound pipes, and numerous related problems. "With a concise and lucid style, clear printing, adequate bibliography and graphs for approximate solutions at the project stage, it fills a vacant place in waterhammer literature," WATER POWER. 43 problems. Bibliography. Index. 113 illustrations. xiv + 161pp. 5⅜ x 8½. S1061 Paperbound **$1.65**

AERODYNAMIC THEORY: A GENERAL REVIEW OF PROGRESS, William F. Durand, editor-in-chief. A monumental joint effort by the world's leading authorities prepared under a grant of the Guggenheim Fund for the Promotion of Aeronautics. Intended to provide the student and aeronautic designer with the theoretical and experimental background of aeronautics. Never equalled for breadth, depth, reliability. Contains discussions of special mathematical topics not usually taught in the engineering or technical courses. Also: an extended two-part treatise on Fluid Mechanics, discussions of aerodynamics of perfect fluids, analyses of experiments with wind tunnels, applied airfoil theory, the non-lifting system of the airplane, the air propeller, hydrodynamics of boats and floats, the aerodynamics of cooling, etc. Contributing experts include Munk, Giacomelli, Prandtl, Toussaint, Von Karman, Klemperer, among others. Unabridged republication. 6 volumes bound as 3. Total of 1,012 figures, 12 plates. Total of 2,186pp. Bibliographies. Notes. Indices. 5⅜ x 8. S328-S330 Clothbound, The Set **$17.50**

APPLIED HYDRO- AND AEROMECHANICS, L. Prandtl and O. G. Tietjens. Presents, for the most part, methods which will be valuable to engineers. Covers flow in pipes, boundary layers, airfoil theory, entry conditions, turbulent flow in pipes, and the boundary layer, determining drag from measurements of pressure and velocity, etc. "Will be welcomed by all students of aerodynamics," NATURE. Unabridged, unaltered. An Engineering Society Monograph, 1934. Index. 226 figures, 28 photographic plates illustrating flow patterns. xvi + 311pp. 5⅜ x 8. S375 Paperbound **$1.85**

SUPERSONIC AERODYNAMICS, E. R. C. Miles. Valuable theoretical introduction to the supersonic domain, with emphasis on mathematical tools and principles, for practicing aerodynamicists and advanced students in aeronautical engineering. Covers fundamental theory, divergence theorem and principles of circulation, compressible flow and Helmholtz laws, the Prandtl-Busemann graphic method for 2-dimensional flow, oblique shock waves, the Taylor-Maccoll method for cones in supersonic flow, the Chaplygin method for 2-dimensional flow, etc. Problems range from practical engineering problems to development of theoretical results. "Rendered outstanding by the unprecedented scope of its contents . . . has undoubtedly filled a vital gap," AERONAUTICAL ENGINEERING REVIEW. Index. 173 problems, answers. 106 diagrams. 7 tables. xii + 255pp. 5⅜ x 8. S214 Paperbound **$1.45**

HYDRAULIC TRANSIENTS, G. R. Rich. The best text in hydraulics ever printed in English . . . by one of America's foremost engineers (former Chief Design Engineer for T.V.A.). Provides a transition from the basic differential equations of hydraulic transient theory to the arithmetic intergration computation required by practicing engineers. Sections cover Water Hammer, Turbine Speed Regulation, Stability of Governing, Water-Hammer Pressures in Pump Discharge Lines, The Differential and Restricted Orifice Surge Tanks, The Normalized Surge Tank Charts of Calame and Gaden, Navigation Locks, Surges in Power Canals—Tidal Harmonics, etc. Revised and enlarged. Author's prefaces. Index. xiv + 409pp. 5⅜ x 8½. S116 Paperbound **$2.50**

HYDRAULICS AND ITS APPLICATIONS, A. H. Gibson. Excellent comprehensive textbook for the student and thorough practical manual for the professional worker, a work of great stature in its area. Half the book is devoted to theory and half to applications and practical problems met in the field. Covers modes of motion of a fluid, critical velocity, viscous flow, eddy formation, Bernoulli's theorem, flow in converging passages, vortex motion, form of effluent streams, notches and weirs, skin friction, losses at valves and elbows, siphons, erosion of channels, jet propulsion, waves of oscillation, and over 100 similar topics. Final chapters (nearly 400 pages) cover more than 100 kinds of hydraulic machinery: Pelton wheel, speed regulators, the hydraulic ram, surge tanks, the scoop wheel, the Venturi meter, etc. A special chapter treats methods of testing theoretical hypotheses: scale models of rivers, tidal estuaries, siphon spillways, etc. 5th revised and enlarged (1952) edition. Index. Appendix. 427 photographs and diagrams. 95 examples, answers. xv + 813pp. 6 x 9. S791 Clothbound **$8.00**

FLUID MECHANICS THROUGH WORKED EXAMPLES, D. R. L. Smith and J. Houghton. Advanced text covering principles and applications to practical situations. Each chapter begins with concise summaries of fundamental ideas. 163 fully worked out examples applying principles outlined in the text. 275 other problems, with answers. Contents; The Pressure of Liquids on Surfaces; Floating Bodies; Flow Under Constant Head in Pipes; Circulation; Vorticity; The Potential Function; Laminar Flow and Lubrication; Impact ot Jets; Hydraulic Turbines; Centrifugal and Reciprocating Pumps; Compressible Fluids; and many other items. Total of 438 examples. 250 line illustrations. 340pp. Index. 6 x 8⅞. S981 Clothbound **$6.00**

THEORY OF SHIP MOTIONS, S. N. Blagoveshchensky. The only detailed text in English in a rapidly developing branch of engineering and physics, it is the work of one of the world's foremost authorities—Blagoveshchensky of Leningrad Shipbuilding Institute. A senior-level treatment written primarily for engineering students, but also of great importance to naval architects, designers, contractors, researchers in hydrodynamics, and other students. No mathematics beyond ordinary differential equations is required for understanding the text. Translated by T. & L. Strelkoff, under editorship of Louis Landweber, Iowa Institute of Hydraulic Research, under auspices of Office of Naval Research. Bibliography. Index. 231 diagrams and illustrations. Total of 649pp. 5⅜ x 8½. Vol. I: S234 Paperbound **$2.00**
Vol. II: S235 Paperbound **$2.00**

THEORY OF FLIGHT, Richard von Mises. Remains almost unsurpassed as balanced, well-written account of fundamental fluid dynamics, and situations in which air compressibility effects are unimportant. Stressing equally theory and practice, avoiding formidable mathematical structure, it conveys a full understanding of physical phenomena and mathematical concepts. Contains perhaps the best introduction to general theory of stability. "Outstanding," *Scientific, Medical, and Technical Books*. New introduction by K. H. Hohenemser. Bibliographical, historical notes. Index. 408 illustrations. xvi + 620pp. 5⅜ x 8⅜. S541 Paperbound **$2.95**

THEORY OF WING SECTIONS, I. H. Abbott, A. E. von Doenhoff. Concise compilation of subsonic aerodynamic characteristics of modern NASA wing sections, with description of their geometry, associated theory. Primarily reference work for engineers, students, it gives methods, data for using wing-section data to predict characteristics. Particularly valuable: chapters on thin wings, airfoils; complete summary of NACA's experimental observations, system of construction families of airfoils. 350pp. of tables on Basic Thickness Forms, Mean Lines, Airfoil Ordinates, Aerodynamic Characteristics of Wing Sections. Index. Bibliography. 191 illustrations. Appendix. 705pp. 5⅜ x 8. S558 Paperbound **$3.25**

WEIGHT-STRENGTH ANALYSIS OF AIRCRAFT STRUCTURES, F. R. Shanley. Scientifically sound methods of analyzing and predicting the structural weight of aircraft and missiles. Deals directly with forces and the distances over which they must be transmitted, making it possible to develop methods by which the minimum structural weight can be determined for any material and conditions of loading. Weight equations for wing and fuselage structures. Includes author's original papers on inelastic buckling and creep buckling. "Particularly successful in presenting his analytical methods for investigating various optimum design principles," AERONAUTICAL ENGINEERING REVIEW. Enlarged bibliography. Index. 199 figures. xiv + 404pp. 5⅝ x 8⅜. S660 Paperbound **$2.50**

Electricity

TWO-DIMENSIONAL FIELDS IN ELECTRICAL ENGINEERING, L. V. Bewley. A useful selection of typical engineering problems of interest to practicing electrical engineers. Introduces senior students to the methods and procedures of mathematical physics. Discusses theory of functions of a complex variable, two-dimensional fields of flow, general theorems of mathematical physics and their applications, conformal mapping or transformation, method of images, freehand flux plotting, etc. New preface by the author. Appendix by W. F. Kiltner. Index. Bibliography at chapter ends. xiv + 204pp. 5⅜ x 8½. S1118 Paperbound **$1.50**

FLUX LINKAGES AND ELECTROMAGNETIC INDUCTION, L. V. Bewley. A brief, clear book which shows proper uses and corrects misconceptions of Faraday's law of electromagnetic induction in specific problems. Contents: Circuits, Turns, and Flux Linkages; Substitution of Circuits; Electromagnetic Induction; General Criteria for Electromagnetic Induction; Applications and Paradoxes; Theorem of Constant Flux Linkages. New Section: Rectangular Coil in a Varying Uniform Medium. Valuable supplement to class texts for engineering students. Corrected, enlarged edition. New preface. Bibliography in notes. 49 figures. xi + 106pp. 5⅜ x 8. S1103 Paperbound **$1.25**

INDUCTANCE CALCULATIONS: WORKING FORMULAS AND TABLES, Frederick W. Grover. An invaluable book to everyone in electrical engineering. Provides simple single formulas to cover all the more important cases of inductance. The approach involves only those parameters that naturally enter into each situation, while extensive tables are given to permit easy interpolations. Will save the engineer and student countless hours and enable them to obtain accurate answers with minimal effort. Corrected republication of 1946 edition. 58 tables. 97 completely worked out examples. 66 figures. xiv + 286pp. 5⅜ x 8½. S974 Paperbound **$1.85**

GASEOUS CONDUCTORS: THEORY AND ENGINEERING APPLICATIONS, J. D. Cobine. An indispensable text and reference to gaseous conduction phenomena, with the engineering viewpoint prevailing throughout. Studies the kinetic theory of gases, ionization, emission phenomena; gas breakdown, spark characteristics, glow, and discharges; engineering applications in circuit interrupters, rectifiers, light sources, etc. Separate detailed treatment of high pressure arcs (Suits); low pressure arcs (Langmuir and Tonks). Much more. "Well organized, clear, straightforward," Tonks, Review of Scientific Instruments. Index. Bibliography. 83 practice problems. 7 appendices. Over 600 figures. 58 tables. xx + 606pp. 5⅜ x 8.   S442 Paperbound **$2.95**

INTRODUCTION TO THE STATISTICAL DYNAMICS OF AUTOMATIC CONTROL SYSTEMS, V. V. Solodovnikov. First English publication of text-reference covering important branch of automatic control systems—random signals; in its original edition, this was the first comprehensive treatment. Examines frequency characteristics, transfer functions, stationary random processes, determination of minimum mean-squared error, of transfer function for a finite period of observation, much more. Translation edited by J. B. Thomas, L. A. Zadeh. Index. Bibliography. Appendix. xxii + 308pp. 5⅜ x 8.   S420 Paperbound **$2.25**

TENSORS FOR CIRCUITS, Gabriel Kron. A boldly original method of analyzing engineering problems, at center of sharp discussion since first introduced, now definitely proved useful in such areas as electrical and structural networks on automatic computers. Encompasses a great variety of specific problems by means of a relatively few symbolic equations. "Power and flexibility . . . becoming more widely recognized," Nature. Formerly "A Short Course in Tensor Analysis." New introduction by B. Hoffmann. Index. Over 800 diagrams. xix + 250pp. 5⅜ x 8.   S534 Paperbound **$2.00**

SELECTED PAPERS ON SEMICONDUCTOR MICROWAVE ELECTRONICS, edited by Sumner N. Levine and Richard R. Kurzrok. An invaluable collection of important papers dealing with one of the most remarkable devolopments in solid-state electronics—the use of the **p-n** junction to achieve amplification and frequency conversion of microwave frequencies. Contents: General Survey (3 introductory papers by W. E. Danielson, R. N. Hall, and M. Tenzer); General Theory of Nonlinear Elements (3 articles by A. van der Ziel, H. E. Rowe, and Manley and Rowe); Device Fabrication and Characterization (3 pieces by Bakanowski, Cranna, and Uhlir, by McCotter, Walker and Fortini, and by S. T. Eng); Parametric Amplifiers and Frequency Multipliers (13 articles by Uhlir, Heffner and Wade, Matthaei, P. K. Tien, van der Ziel, Engelbrecht, Currie and Gould, Uenohara, Leeson and Weinreb, and others); and Tunnel Diodes (4 papers by L. Esaki, H. S. Sommers, Jr., M. E. Hines, and Yariv and Cook). Introduction. 295 Figures. xiii + 286pp. 6½ x 9¼.   S1126 Paperbound **$2.25**

THE PRINCIPLES OF ELECTROMAGNETISM APPLIED TO ELECTRICAL MACHINES, B. Hague. A concise, but complete, summary of the basic principles of the magnetic field and its applications, with particular reference to the kind of phenomena which occur in electrical machines. Part I: General Theory—magnetic field of a current, electromagnetic field passing from air to iron, mechanical forces on linear conductors, etc. Part II: Application of theory to the solution of electromechanical problems—the magnetic field and mechanical forces in non-salient pole machinery, the field within slots and between salient poles, and the work of Rogowski, Roth, and Strutt. Formery titled "Electromagnetic Problems in Electrical Engineering." 2 appendices. Index. Bibliography in notes. 115 figures. xiv + 359pp. 5⅜ x 8½.   S246 Paperbound **$2.25**

Mechanical engineering

DESIGN AND USE OF INSTRUMENTS AND ACCURATE MECHANISM, T. N. Whitehead. For the instrument designer, engineer; how to combine necessary mathematical abstractions with independent observation of actual facts. Partial contents: instruments & their parts, theory of errors, systematic errors, probability, short period errors, erratic errors, design precision, kinematic, semikinematic design, stiffness, planning of an instrument, human factor, etc. Index. 85 photos, diagrams. xii + 288pp. 5⅜ x 8.   S270 Paperbound **$2.00**

A TREATISE ON GYROSTATICS AND ROTATIONAL MOTION: THEORY AND APPLICATIONS, Andrew Gray. Most detailed, thorough book in English, generally considered definitive study. Many problems of all sorts in full detail, or step-by-step summary. Classical problems of Bour, Lottner, etc.; later ones of great physical interest. Vibrating systems of gyrostats, earth as a top, calculation of path of axis of a top by elliptic integrals, motion of unsymmetrical top, much more. Index. 160 illus. 550pp. 5⅜ x 8.   S589 Paperbound **$2.75**

MECHANICS OF THE GYROSCOPE, THE DYNAMICS OF ROTATION, R. F. Deimel, Professor of Mechanical Engineering at Stevens Institute of Technology. Elementary general treatment of dynamics of rotation, with special application of gyroscopic phenomena. No knowledge of vectors needed. Velocity of a moving curve, acceleration to a point, general equations of motion, gyroscopic horizon, free gyro, motion of discs, the damped gyro, 103 similar topics. Exercises. 75 figures. 208pp. 5⅜ x 8.   S66 Paperbound **$1.75**

Optical design, lighting

THE SCIENTIFIC BASIS OF ILLUMINATING ENGINEERING, Parry Moon, Professor of Electrical Engineering, M.I.T. Basic, comprehensive study. Complete coverage of the fundamental theoretical principles together with the elements of design, vision, and color with which the lighting engineer must be familiar. Valuable as a text as well as a reference source to the practicing engineer. Partial contents: Spectroradiometric Curve, Luminous Flux, Radiation from Gaseous-Conduction Sources, Radiation from Incandescent Sources, Incandescent Lamps, Measurement of Light, Illumination from Point Sources and Surface Sources, Elements of Lighting Design. 7 Appendices. Unabridged and corrected republication, with additions. New preface containing conversion tables of radiometric and photometric concepts. Index. 707-item bibliography. 92-item bibliography of author's articles. 183 problems. xxiii + 608pp. 5⅜ x 8½. S242 Paperbound **$2.85**

OPTICS AND OPTICAL INSTRUMENTS: AN INTRODUCTION WITH SPECIAL REFERENCE TO PRACTICAL APPLICATIONS, B. K. Johnson. An invaluable guide to basic practical applications of optical principles, which shows how to set up inexpensive working models of each of the four main types of optical instruments—telescopes, microscopes, photographic lenses, optical projecting systems. Explains in detail the most important experiments for determining their accuracy, resolving power, angular field of view, amounts of aberration, all other necessary facts about the instruments. Formerly "Practical Optics." Index. 234 diagrams. Appendix. 224pp. 5⅜ x 8. S642 Paperbound **$1.65**

APPLIED OPTICS AND OPTICAL DESIGN, A. E. Conrady. With publication of vol. 2, standard work for designers in optics is now complete for first time. Only work of its kind in English; only detailed work for practical designer and self-taught. Requires, for bulk of work, no math above trig. Step-by-step exposition, from fundamental concepts of geometrical, physical optics, to systematic study, design, of almost all types of optical systems. Vol. 1: all ordinary ray-tracing methods; primary aberrations; necessary higher aberration for design of telescopes, low-power microscopes, photographic equipment. Vol. 2: (Completed from author's notes by R. Kingslake, Dir. Optical Design, Eastman Kodak.) Special attention to high-power microscope, anastigmatic photographic objectives. "An indispensable work," J., Optical Soc. of Amer. "As a practical guide this book has no rival," Transactions, Optical Soc. Index. Bibliography. 193 diagrams. 852pp. 6⅛ x 9¼. Vol. 1 S366 Paperbound **$3.50** Vol. 2 S612 Paperbound **$2.95**

Miscellaneous

THE MEASUREMENT OF POWER SPECTRA FROM THE POINT OF VIEW OF COMMUNICATIONS ENGINEERING, R. B. Blackman, J. W. Tukey. This pathfinding work, reprinted from the "Bell System Technical Journal," explains various ways of getting practically useful answers in the measurement of power spectra, using results from both transmission theory and the theory of statistical estimation. Treats: Autocovariance Functions and Power Spectra; Direct Analog Computation; Distortion, Noise, Heterodyne Filtering and Pre-whitening; Aliasing; Rejection Filtering and Separation; Smoothing and Decimation Procedures; Very Low Frequencies; Transversal Filtering; much more. An appendix reviews fundamental Fourier techniques. Index of notation. Glossary of terms. 24 figures. XII tables. Bibliography. General index. 192pp. 5⅜ x 8. S507 Paperbound **$1.85**

CALCULUS REFRESHER FOR TECHNICAL MEN, A. Albert Klaf. This book is unique in English as a refresher for engineers, technicians, students who either wish to brush up their calculus or to clear up uncertainties. It is not an ordinary text, but an examination of most important aspects of integral and differential calculus in terms of the 756 questions most likely to occur to the technical reader. The first part of this book covers simple differential calculus, with constants, variables, functions, increments, derivatives, differentiation, logarithms, curvature of curves, and similar topics. The second part covers fundamental ideas of integration, inspection, substitution, transformation, reduction, areas and volumes, mean value, successive and partial integration, double and triple integration. Practical aspects are stressed rather than theoretical. A 50-page section illustrates the application of calculus to specific problems of civil and nautical engineering, electricity, stress and strain, elasticity, industrial engineering, and similar fields.—756 questions answered. 566 problems, mostly answered. 36 pages of useful constants, formulae for ready reference. Index. v + 431pp. 5⅜ x 8. T370 Paperbound **$2.00**

METHODS IN EXTERIOR BALLISTICS, Forest Ray Moulton. Probably the best introduction to the mathematics of projectile motion. The ballistics theories propounded were coordinated with extensive proving ground and wind tunnel experiments conducted by the author and others for the U.S. Army. Broad in scope and clear in exposition, it gives the beginnings of the theory used for modern-day projectile, long-range missile, and satellite motion. Six main divisions: Differential Equations of Translatory Motion of a projectile; Gravity and the Resistance Function; Numerical Solution of Differential Equations; Theory of Differential Variations; Validity of Method of Numerical Integration; and Motion of a Rotating Projectile. Formerly titled: "New Methods in Exterior Ballistics." Index. 38 diagrams. viii + 259pp. 5⅜ x 8½. S232 Paperbound **$1.75**

LOUD SPEAKERS: THEORY, PERFORMANCE, TESTING AND DESIGN, N. W. McLachlan. Most comprehensive coverage of theory, practice of loud speaker design, testing; classic reference, study manual in field. First 12 chapters deal with theory, for readers mainly concerned with math. aspects; last 7 chapters will interest reader concerned with testing, design. Partial contents: principles of sound propagation, fluid pressure on vibrators, theory of moving-coil principle, transients, driving mechanisms, response curves, design of horn type moving coil speakers, electrostatic speakers, much more. Appendix. Bibliography. Index. 165 illustrations, charts. 411pp. 5⅜ x 8.
S588 Paperbound **$2.25**

MICROWAVE TRANSMISSION, J. C. Slater. First text dealing exclusively with microwaves, brings together points of view of field, circuit theory, for graduate student in physics, electrical engineering, microwave technician. Offers valuable point of view not in most later studies. Uses Maxwell's equations to study electromagnetic field, important in this area. Partial contents: infinite line with distributed parameters, impedance of terminated line, plane waves, reflections, wave guides, coaxial line, composite transmission lines, impedance matching, etc. Introduction. Index. 76 illus. 319pp. 5⅜ x 8.
S564 Paperbound **$1.50**

MICROWAVE TRANSMISSION DESIGN DATA, T. Moreno. Originally classified, now rewritten and enlarged (14 new chapters) for public release under auspices of Sperry Corp. Material of immediate value or reference use to radio engineers, systems designers, applied physicists, etc. Ordinary transmission line theory; attenuation; capacity; parameters of coaxial lines; higher modes; flexible cables; obstacles, discontinuities, and injunctions; tunable wave guide impedance transformers; effects of temperature and humidity; much more. "Enough theoretical discussion is included to allow use of data without previous background," Electronics. 324 circuit diagrams, figures, etc. Tables of dielectrics, flexible cable, etc., data. Index. ix + 248pp. 5⅜ x 8.
S459 Paperbound **$1.65**

RAYLEIGH'S PRINCIPLE AND ITS APPLICATIONS TO ENGINEERING, G. Temple & W. Bickley. Rayleigh's principle developed to provide upper and lower estimates of true value of fundamental period of a vibrating system, or condition of stability of elastic systems. Illustrative examples; rigorous proofs in special chapters. Partial contents: Energy method of discussing vibrations, stability. Perturbation theory, whirling of uniform shafts. Criteria of elastic stability. Application of energy method. Vibrating systems. Proof, accuracy, successive approximations, application of Rayleigh's principle. Synthetic theorems. Numerical, graphical methods. Equilibrium configurations, Ritz's method. Bibliography. Index. 22 figures. ix + 156pp. 5⅜ x 8.
S307 Paperbound **$1.50**

ELASTICITY, PLASTICITY AND STRUCTURE OF MATTER, R. Houwink. Standard treatise on rheological aspects of different technically important solids such as crystals, resins, textiles, rubber, clay, many others. Investigates general laws for deformations; determines divergences from these laws for certain substances. Covers general physical and mathematical aspects of plasticity, elasticity, viscosity. Detailed examination of deformations, internal structure of matter in relation to elastic and plastic behavior, formation of solid matter from a fluid, conditions for elastic and plastic behavior of matter. Treats glass, asphalt, gutta percha, balata, proteins, baker's dough, lacquers, sulphur, others. 2nd revised, enlarged edition. Extensive revised bibliography in over 500 footnotes. Index. Table of symbols. 214 figures. xviii + 368pp. 6 x 9¼.
S385 Paperbound **$2.45**

THE SCHWARZ-CHRISTOFFEL TRANSFORMATION AND ITS APPLICATIONS: A SIMPLE EXPOSITION, Miles Walker. An important book for engineers showing how this valuable tool can be employed in practical situations. Very careful, clear presentation covering numerous concrete engineering problems. Includes a thorough account of conjugate functions for engineers—useful for the beginner and for review. Applications to such problems as: Stream-lines round a corner, electric conductor in air-gap, dynamo slots, magnetized poles, much more. Formerly "Conjugate Functions for Engineers." Preface. 92 figures, several tables. Index. ix + 116pp. 5⅜ x 8½.
S1149 Paperbound **$1.25**

THE LAWS OF THOUGHT, George Boole. This book founded symbolic logic some hundred years ago. It is the 1st significant attempt to apply logic to all aspects of human endeavour. Partial contents: derivation of laws, signs & laws, interpretations, eliminations, conditions of a perfect method, analysis, Aristotelian logic, probability, and similar topics. xviii + 424pp. 5⅜ x 8.
S28 Paperbound **$2.00**

SCIENCE AND METHOD, Henri Poincaré. Procedure of scientific discovery, methodology, experiment, idea-germination—the intellectual processes by which discoveries come into being. Most significant and most interesting aspects of development, application of ideas. Chapters cover selection of facts, chance, mathematical reasoning, mathematics, and logic; Whitehead, Russell, Cantor; the new mechanics, etc. 288pp. 5⅜ x 8.
S222 Paperbound **$1.35**

FAMOUS BRIDGES OF THE WORLD, D. B. Steinman. An up-to-the-minute revised edition of a book that explains the fascinating drama of how the world's great bridges came to be built. The author, designer of the famed Mackinac bridge, discusses bridges from all periods and all parts of the world, explaining their various types of construction, and describing the problems their builders faced. Although primarily for youngsters, this cannot fail to interest readers of all ages. 48 illustrations in the text. 23 photographs. 99pp. 6⅛ x 9¼.
T161 Paperbound **$1.00**

PHYSICS

General physics

FOUNDATIONS OF PHYSICS, R. B. Lindsay & H. Margenau. Excellent bridge between semi-popular works & technical treatises. A discussion of methods of physical description, construction of theory; valuable for physicist with elementary calculus who is interested in ideas that give meaning to data, tools of modern physics. Contents include symbolism, mathematical equations; space & time foundations of mechanics; probability; physics & continua; electron theory; special & general relativity; quantum mechanics; causality. "Thorough and yet not overdetailed. Unreservedly recommended," NATURE (London). Unabridged, corrected edition. List of recommended readings. 35 illustrations. xi + 537pp. 5⅜ x 8.
S377 Paperbound **$2.75**

FUNDAMENTAL FORMULAS OF PHYSICS, ed. by D. H. Menzel. Highly useful, fully inexpensive reference and study text, ranging from simple to highly sophisticated operations. Mathematics integrated into text—each chapter stands as short textbook of field represented. Vol. 1: Statistics, Physical Constants, Special Theory of Relativity, Hydrodynamics, Aerodynamics, Boundary Value Problems in Math. Physics; Viscosity, Electromagnetic Theory, etc. Vol. 2: Sound, Acoustics, Geometrical Optics, Electron Optics, High-Energy Phenomena, Magnetism, Biophysics, much more. Index. Total of 800pp. 5⅜ x 8.   Vol. 1 S595 Paperbound **$2.00**
Vol. 2 S596 Paperbound **$2.00**

MATHEMATICAL PHYSICS, D. H. Menzel. Thorough one-volume treatment of the mathematical techniques vital for classic mechanics, electromagnetic theory, quantum theory, and relativity. Written by the Harvard Professor of Astrophysics for junior, senior, and graduate courses, it gives clear explanations of all those aspects of function theory, vectors, matrices, dyadics, tensors, partial differential equations, etc., necessary for the understanding of the various physical theories. Electron theory, relativity, and other topics seldom presented appear here in considerable detail. Scores of definitions, conversion factors, dimensional constants, etc. "More detailed than normal for an advanced text . . . excellent set of sections on Dyadics, Matrices, and Tensors," JOURNAL OF THE FRANKLIN INSTITUTE. Index. 193 problems, with answers. x + 412pp. 5⅜ x 8.   S56 Paperbound **$2.00**

THE SCIENTIFIC PAPERS OF J. WILLARD GIBBS. All the published papers of America's outstanding theoretical scientist (except for "Statistical Mechanics" and "Vector Analysis"). Vol I (thermodynamics) contains one of the most brilliant of all 19th-century scientific papers—the 300-page "On the Equilibrium of Heterogeneous Substances," which founded the science of physical chemistry, and clearly stated a number of highly important natural laws for the first time; 8 other papers complete the first volume. Vol II includes 2 papers on dynamics, 8 on vector analysis and multiple algebra, 5 on the electromagnetic theory of light, and 6 miscellaneous papers. Biographical sketch by H. A. Bumstead. Total of xxxvi + 718pp. 5⅝ x 8⅜.
S721 Vol I Paperbound **$2.50**
S722 Vol II Paperbound **$2.00**
The set **$4.50**

BASIC THEORIES OF PHYSICS, Peter Gabriel Bergmann. Two-volume set which presents a critical examination of important topics in the major subdivisions of classical and modern physics. The first volume is concerned with classical mechanics and electrodynamics: mechanics of mass points, analytical mechanics, matter in bulk, electrostatics and magnetostatics, electromagnetic interaction, the field waves, special relativity, and waves. The second volume (Heat and Quanta) contains discussions of the kinetic hypothesis, physics and statistics, stationary ensembles, laws of thermodynamics, early quantum theories, atomic spectra, probability waves, quantization in wave mechanics, approximation methods, and abstract quantum theory. A valuable supplement to any thorough course or text.
Heat and Quanta: Index. 8 figures. x + 300pp. 5⅜ x 8½.   S968 Paperbound **$1.75**
Mechanics and Electrodynamics: Index. 14 figures. vii + 280pp. 5⅜ x 8½.
S969 Paperbound **$1.75**

THEORETICAL PHYSICS, A. S. Kompaneyets. One of the very few thorough studies of the subject in this price range. Provides advanced students with a comprehensive theoretical background. Especially strong on recent experimentation and developments in quantum theory. Contents: Mechanics (Generalized Coordinates, Lagrange's Equation, Collision of Particles, etc.), Electrodynamics (Vector Analysis, Maxwell's equations, Transmission of Signals, Theory of Relativity, etc.), Quantum Mechanics (the Inadequacy of Classical Mechanics, the Wave Equation, Motion in a Central Field, Quantum Theory of Radiation, Quantum Theories of Dispersion and Scattering, etc.), and Statistical Physics (Equilibrium Distribution of Molecules in an Ideal Gas, Boltzmann statistics, Bose and Fermi Distribution, Thermodynamic Quantities, etc.). Revised to 1961. Translated by George Yankovsky, authorized by Kompaneyets. 137 exercises. 56 figures. 529pp. 5⅜ x 8½. S972 Paperbound **$2.50**

ANALYTICAL AND CANONICAL FORMALISM IN PHYSICS, André Mercier. A survey, in one volume, of the variational principles (the key principles—in mathematical form—from which the basic laws of any one branch of physics can be derived) of the several branches of physical theory, together with an examination of the relationships among them. Contents: the Lagrangian Formalism, Lagrangian Densities, Canonical Formalism, Canonical Form of Electrodynamics, Hamiltonian Densities, Transformations, and Canonical Form with Vanishing Jacobian Determinant. Numerous examples and exercises. For advanced students, teachers, etc. 6 figures. Index. viii + 222pp. 5⅜ x 8½.   S1077 Paperbound **$1.75**

Acoustics, optics, electricity and magnetism, electromagnetics, magneto-hydrodynamics

THE THEORY OF SOUND, Lord Rayleigh. Most vibrating systems likely to be encountered in practice can be tackled successfully by the methods set forth by the great Nobel laureate, Lord Rayleigh. Complete coverage of experimental, mathematical aspects of sound theory. Partial contents: Harmonic motions, vibrating systems in general, lateral vibrations of bars, curved plates or shells, applications of Laplace's functions to acoustical problems, fluid friction, plane vortex-sheet, vibrations of solid bodies, etc. This is the first inexpensive edition of this great reference and study work. Bibliography. Historical introduction by R. B. Lindsay. Total of 1040pp. 97 figures. 5⅜ x 8.
S292, S293, Two volume set, paperbound, **$4.70**

THE DYNAMICAL THEORY OF SOUND, H. Lamb. Comprehensive mathematical treatment of the physical aspects of sound, covering the theory of vibrations, the general theory of sound, and the equations of motion of strings, bars, membranes, pipes, and resonators. Includes chapters on plane, spherical, and simple harmonic waves, and the Helmholtz Theory of Audition. Complete and self-contained development for student and specialist; all fundamental differential equations solved completely. Specific mathematical details for such important phenomena as harmonics, normal modes, forced vibrations of strings, theory of reed pipes, etc. Index. Bibliography. 86 diagrams. viii + 307pp. 5⅜ x 8.
S655 Paperbound **$1.50**

WAVE PROPAGATION IN PERIODIC STRUCTURES, L. Brillouin. A general method and application to different problems: pure physics, such as scattering of X-rays of crystals, thermal vibration in crystal lattices, electronic motion in metals; and also problems of electrical engineering. Partial contents: elastic waves in 1-dimensional lattices of point masses. Propagation of waves along 1-dimensional lattices. Energy flow. 2 dimensional, 3 dimensional lattices. Mathieu's equation. Matrices and propagation of waves along an electric line. Continuous electric lines. 131 illustrations. Bibliography. Index. xii + 253pp. 5⅜ x 8.
S34 Paperbound **$2.00**

THEORY OF VIBRATIONS, N. W. McLachlan. Based on an exceptionally successful graduate course given at Brown University, this discusses linear systems having 1 degree of freedom, forced vibrations of simple linear systems, vibration of flexible strings, transverse vibrations of bars and tubes, transverse vibration of circular plate, sound waves of finite amplitude, etc. Index. 99 diagrams. 160pp. 5⅜ x 8.
S190 Paperbound **$1.35**

LIGHT: PRINCIPLES AND EXPERIMENTS, George S. Monk. Covers theory, experimentation, and research. Intended for students with some background in general physics and elementary calculus. Three main divisions: 1) Eight chapters on geometrical optics—fundamental concepts (the ray and its optical length, Fermat's principle, etc.), laws of image formation, apertures in optical systems, photometry, optical instruments etc.; 2) 9 chapters on physical optics—interference, diffraction, polarization, spectra, the Rayleigh refractometer, the wave theory of light, etc.; 3) 23 instructive experiments based directly on the theoretical text. "Probably the best intermediate textbook on light in the English language. Certainly, it is the best book which includes both geometrical and physical optics," J. Rud Nielson, PHYSICS FORUM. Revised edition. 102 problems and answers. 12 appendices. 6 tables. Index. 270 illustrations. xi +489pp. 5⅜ x 8½.
S341 Paperbound **$2.50**

PHOTOMETRY, John W. T. Walsh. The best treatment of both "bench" and "illumination" photometry in English by one of Britain's foremost experts in the field (President of the International Commission on Illumination). Limited to those matters, theoretical and practical, which affect the measurement of light flux, candlepower, illumination, etc., and excludes treatment of the use to which such measurements may be put after they have been made. Chapters on Radiation, The Eye and Vision, Photo-Electric Cells, The Principles of Photometry, The Measurement of Luminous Intensity, Colorimetry, Spectrophotometry, Stellar Photometry, The Photometric Laboratory, etc. Third revised (1958) edition. 281 illustrations. 10 appendices. xxiv + 544pp. 5½ x 9¼.
S319 Clothbound **$10.00**

EXPERIMENTAL SPECTROSCOPY, R. A. Sawyer. Clear discussion of prism and grating spectrographs and the techniques of their use in research, with emphasis on those principles and techniques that are fundamental to practically all uses of spectroscopic equipment. Beginning with a brief history of spectroscopy, the author covers such topics as light sources, spectroscopic apparatus, prism spectroscopes and gratings, diffraction grating, the photographic process, determination of wave length, spectral intensity, infrared spectroscopy, spectrochemical analysis, etc. This revised edition contains new material on the production of replica gratings, solar spectroscopy from rockets, new standard of wave length, etc. Index. Bibliography. 111 illustrations. x + 358pp. 5⅜ x 8½.
S1045 Paperbound **$2.25**

FUNDAMENTALS OF ELECTRICITY AND MAGNETISM, L. B. Loeb. For students of physics, chemistry, or engineering who want an introduction to electricity and magnetism on a higher level and in more detail than general elementary physics texts provide. Only elementary differential and integral calculus is assumed. Physical laws developed logically, from magnetism to electric currents, Ohm's law, electrolysis, and on to static electricity, induction, etc. Covers an unusual amount of material; one third of book on modern material: solution of wave equation, photoelectric and thermionic effects, etc. Complete statement of the various electrical systems of units and interrelations. 2 Indexes. 75 pages of problems with answers stated. Over 300 figures and diagrams. xix +669pp. 5⅜ x 8.
S745 Paperbound **$2.75**

MATHEMATICAL ANALYSIS OF ELECTRICAL AND OPTICAL WAVE-MOTION, Harry Bateman. Written by one of this century's most distinguished mathematical physicists, this is a practical introduction to those developments of Maxwell's electromagnetic theory which are directly connected with the solution of the partial differential equation of wave motion. Methods of solving wave-equation, polar-cylindrical coordinates, diffraction, transformation of coordinates, homogeneous solutions, electromagnetic fields with moving singularities, etc. Index. 168pp. 5⅜ x 8. S14 Paperbound **$1.75**

PRINCIPLES OF PHYSICAL OPTICS, Ernst Mach. This classical examination of the propagation of light, color, polarization, etc. offers an historical and philosophical treatment that has never been surpassed for breadth and easy readability. Contents: Rectilinear propagation of light. Reflection, refraction. Early knowledge of vision. Dioptrics. Composition of light. Theory of color and dispersion. Periodicity. Theory of interference. Polarization. Mathematical representation of properties of light. Propagation of waves, etc. 279 illustrations, 10 portraits. Appendix. Indexes. 324pp. 5⅜ x 8. S178 Paperbound **$2.00**

THE THEORY OF OPTICS, Paul Drude. One of finest fundamental texts in physical optics, classic offers thorough coverage, complete mathematical treatment of basic ideas. Includes fullest treatment of application of thermodynamics to optics; sine law in formation of images, transparent crystals, magnetically active substances, velocity of light, apertures, effects depending upon them, polarization, optical instruments, etc. Introduction by A. A. Michelson. Index. 110 illus. 567pp. 5⅜ x 8. S532 Paperbound **$2.45**

ELECTRICAL THEORY ON THE GIORGI SYSTEM, P. Cornelius. A new clarification of the fundamental concepts of electricity and magnetism, advocating the convenient m.k.s. system of units that is steadily gaining followers in the sciences. Illustrating the use and effectiveness of his terminology with numerous applications to concrete technical problems, the author here expounds the famous Giorgi system of electrical physics. His lucid presentation and well-reasoned, cogent argument for the universal adoption of this system form one of the finest pieces of scientific exposition in recent years. 28 figures. Index. Conversion tables for translating earlier data into modern units. Translated from 3rd Dutch edition by L. J. Jolley. x + 187pp. 5½ x 8¾. S909 Clothbound **$6.00**

ELECTRIC WAVES: BEING RESEARCHES ON THE PROPAGATION OF ELECTRIC ACTION WITH FINITE VELOCITY THROUGH SPACE, Heinrich Hertz. This classic work brings together the original papers in which Hertz—Helmholtz's protegé and one of the most brilliant figures in 19th-century research—probed the existence of electromagnetic waves and showed experimentally that their velocity equalled that of light, research that helped lay the groundwork for the development of radio, television, telephone, telegraph, and other modern technological marvels. Unabridged republication of original edition. Authorized translation by D. E. Jones. Preface by Lord Kelvin. Index of names. 40 illustrations. xvii + 278pp. 5⅜ x 8½. S57 Paperbound **$1.75**

PIEZOELECTRICITY: AN INTRODUCTION TO THE THEORY AND APPLICATIONS OF ELECTRO-MECHANICAL PHENOMENA IN CRYSTALS, Walter G. Cady. This is the most complete and systematic coverage of this important field in print—now regarded as something of scientific classic. This republication, revised and corrected by Prof. Cady—one of the foremost contributors in this area—contains a sketch of recent progress and new material on Ferroelectrics. Time Standards, etc. The first 7 chapters deal with fundamental theory of crystal electricity. 5 important chapters cover basic concepts of piezoelectricity, including comparisons of various competing theories in the field. Also discussed: piezoelectric resonators (theory, methods of manufacture, influences of air-gaps, etc.); the piezo oscillator; the properties, history, and observations relating to Rochelle salt; ferroelectric crystals; miscellaneous applications of piezoelectricity; pyroelectricity; etc. "A great work," W. A. Wooster, NATURE. Revised (1963) and corrected edition. New preface by Prof. Cady. 2 Appendices. Indices. Illustrations. 62 tables. Bibliography. Problems. Total of 1 + 822pp. 5⅜ x 8½.
S1094 Vol. I Paperbound **$2.50**
S1095 Vol. II Paperbound **$2.50**
Two volume set Paperbound **$5.00**

MAGNETISM AND VERY LOW TEMPERATURES, H. B. G. Casimir. A basic work in the literature of low temperature physics. Presents a concise survey of fundamental theoretical principles, and also points out promising lines of investigation. Contents: Classical Theory and Experimental Methods, Quantum Theory of Paramagnetism, Experiments on Adiabatic Demagnetization. Theoretical Discussion of Paramagnetism at Very Low Temperatures, Some Experimental Results, Relaxation Phenomena. Index. 89-item bibliography. ix + 95pp. 5⅜ x 8. S943 Paperbound **$1.25**

SELECTED PAPERS ON NEW TECHNIQUES FOR ENERGY CONVERSION: THERMOELECTRIC METHODS; THERMIONIC; PHOTOVOLTAIC AND ELECTRICAL EFFECTS; FUSION, Edited by Sumner N. Levine. Brings together in one volume the most important papers (1954-1961) in modern energy technology. Included among the 37 papers are general and qualitative descriptions of the field as a whole, introducing promising lines of research. Also: 15 papers on thermoelectric methods, 7 on thermionic, 5 on photovoltaic, 4 on electrochemical effect, and 2 on controlled fusion research. Among the contributors are: Joffe, Maria Telkes, Herold, Herring, Douglas, Jaumot, Post, Austin, Wilson, Pfann, Rappaport, Morehouse, Domenicali, Moss, Bowers, Harman, Von Doenhoef. Preface and introduction by the editor. Bibliographies. xxviii + 451pp. 6⅛ x 9¼. S37 Paperbound **$3.00**

SUPERFLUIDS: MACROSCOPIC THEORY OF SUPERCONDUCTIVITY, Vol. I, Fritz London. The major work by one of the founders and great theoreticians of modern quantum physics. Consolidates the researches that led to the present understanding of the nature of superconductivity. Prof. London here reveals that quantum mechanics is operative on the macroscopic plane as well as the submolecular level. Contents: Properties of Superconductors and Their Thermodynamical Correlation; Electrodynamics of the Pure Superconducting State; Relation between Current and Field; Measurements of the Penetration Depth; Non-Viscous Flow vs. Superconductivity; Micro-waves in Superconductors; Reality of the Domain Structure; and many other related topics. A new epilogue by M. J. Buckingham discusses developments in the field up to 1960. Corrected and expanded edition. An appreciation of the author's life and work by L. W. Nordheim. Biography by Edith London. Bibliography of his publications. 45 figures. 2 Indices. xviii + 173pp. 5⅝ x 8⅜.   S44 Paperbound **$1.45**

SELECTED PAPERS ON PHYSICAL PROCESSES IN IONIZED PLASMAS, Edited by Donald H. Menzel, Director, Harvard College Observatory. 30 important papers relating to the study of highly ionized gases or plasmas selected by a foremost contributor in the field, with the assistance of Dr. L. H. Aller. The essays include 18 on the physical processes in gaseous nebulae, covering problems of radiation and radiative transfer, the Balmer decrement, electron temperatures, spectrophotometry, etc. 10 papers deal with the interpretation of nebular spectra, by Bohm, Van Vleck, Aller, Minkowski, etc. There is also a discussion of the intensities of "forbidden" spectral lines by George Shortley and a paper concerning the theory of hydrogenic spectra by Menzel and Pekeris. Other contributors: Goldberg, Hebb, Baker, Bowen, Ufford, Liller, etc. viii + 374pp. 6⅛ x 9¼.   S60 Paperbound **$2.95**

THE ELECTROMAGNETIC FIELD, Max Mason & Warren Weaver. Used constantly by graduate engineers. Vector methods exclusively: detailed treatment of electrostatics, expansion methods, with tables converting any quantity into absolute electromagnetic, absolute electrostatic, practical units. Discrete charges, ponderable bodies, Maxwell field equations, etc. Introduction. Indexes. 416pp. 5⅜ x 8.   S185 Paperbound **$2.00**

THEORY OF ELECTRONS AND ITS APPLICATION TO THE PHENOMENA OF LIGHT AND RADIANT HEAT, H. Lorentz. Lectures delivered at Columbia University by Nobel laureate Lorentz. Unabridged, they form a historical coverage of the theory of free electrons, motion, absorption of heat, Zeeman effect, propagation of light in molecular bodies, inverse Zeeman effect, optical phenomena in moving bodies, etc. 109 pages of notes explain the more advanced sections. Index. 9 figures. 352pp. 5⅜ x 8.   S173 Paperbound **$1.85**

FUNDAMENTAL ELECTROMAGNETIC THEORY, Ronold P. King, Professor Applied Physics, Harvard University. Original and valuable introduction to electromagnetic theory and to circuit theory from the standpoint of electromagnetic theory. Contents: Mathematical Description of Matter—stationary and nonstationary states; Mathematical Description of Space and of Simple Media—Field Equations, Integral Forms of Field Equations, Electromagnetic Force, etc.; Transformation of Field and Force Equations; Electromagnetic Waves in Unbounded Regions; Skin Effect and Internal Impedance—in a solid cylindrical conductor, etc.; and Electrical Circuits—Analytical Foundations, Near-zone and quasi-near zone circuits, Balanced two-wire and four-wire transmission lines. Revised and enlarged version. New preface by the author. 5 appendices (Differential operators: Vector Formulas and Identities, etc.). Problems. Indexes. Bibliography. xvi + 580pp. 5⅜ x 8½.   S1023 Paperbound **$2.75**

Hydrodynamics

A TREATISE ON HYDRODYNAMICS, A. B. Basset. Favorite text on hydrodynamics for 2 generations of physicists, hydrodynamical engineers, oceanographers, ship designers, etc. Clear enough for the beginning student, and thorough source for graduate students and engineers on the work of d'Alembert, Euler, Laplace, Lagrange, Poisson, Green, Clebsch, Stokes, Cauchy, Helmholtz, J. J. Thomson, Love, Hicks, Greenhill, Besant, Lamb, etc. Great amount of documentation on entire theory of classical hydrodynamics. Vol I: theory of motion of frictionless liquids, vortex, and cyclic irrotational motion, etc. 132 exercises. Bibliography. 3 Appendixes. xii + 264pp. Vol II: motion in viscous liquids, harmonic analysis, theory of tides, etc. 112 exercises, Bibliography. 4 Appendixes. xv + 328pp. Two volume set. 5⅜ x 8.
'S724 Vol I Paperbound **$1.75**
$725 Vol II Paperbound **$1.75**
The set **$3.50**

HYDRODYNAMICS, Horace Lamb. Internationally famous complete coverage of standard reference work on dynamics of liquids &. gases. Fundamental theorems, equations, methods, solutions, background, for classical hydrodynamics. Chapters include Equations of Motion, Integration of Equations in Special Gases, Irrotational Motion, Motion of Liquid in 2 Dimensions, Motion of Solids through Liquid-Dynamical Theory, Vortex Motion, Tidal Waves, Surface Waves, Waves of Expansion, Viscosity, Rotating Masses of liquids. Excellently planned, arranged; clear, lucid presentation. 6th enlarged, revised edition. Index. Over 900 footnotes, mostly bibliographical. 119 figures. xv + 738pp. 6⅛ x 9¼.   S256 Paperbound **$3.75**

HYDRODYNAMICS, H. Dryden, F. Murnaghan, Harry Bateman. Published by the National Research Council in 1932 this enormous volume offers a complete coverage of classical hydrodynamics. Encyclopedic in quality. Partial contents: physics of fluids, motion, turbulent flow, compressible fluids, motion in 1, 2, 3 dimensions; viscous fluids rotating, laminar motion, resistance of motion through viscous fluid, eddy viscosity, hydraulic flow in channels of various shapes, discharge of gases, flow past obstacles, etc. Bibliography of over 2,900 items. Indexes. 23 figures. 634pp. 5⅜ x 8.   S303 Paperbound $2.75

Mechanics, dynamics, thermodynamics, elasticity

MECHANICS, J. P. Den Hartog. Already a classic among introductory texts, the M.I.T. professor's lively and discursive presentation is equally valuable as a beginner's text, an engineering student's refresher, or a practicing engineer's reference. Emphasis in this highly readable text is on illuminating fundamental principles and showing how they are embodied in a great number of real engineering and design problems: trusses, loaded cables, beams, jacks, hoists, etc. Provides advanced material on relative motion and gyroscopes not usual in introductory texts. "Very thoroughly recommended to all those anxious to improve their real understanding of the principles of mechanics." MECHANICAL WORLD. Index. List of equations. 334 problems, all with answers. Over 550 diagrams and drawings. ix + 462pp. 5⅜ x 8.
S754 Paperbound $2.00

THEORETICAL MECHANICS: AN INTRODUCTION TO MATHEMATICAL PHYSICS, J. S. Ames, F. D. Murnaghan. A mathematically rigorous development of theoretical mechanics for the advanced student, with constant practical applications. Used in hundreds of advanced courses. An unusually thorough coverage of gyroscopic and baryscopic material, detailed analyses of the Coriolis acceleration, applications of Lagrange's equations, motion of the double pendulum, Hamilton-Jacobi partial differential equations, group velocity and dispersion, etc. Special relativity is also included. 159 problems. 44 figures. ix + 462pp. 5⅜ x 8.
S461 Paperbound $2.25

THEORETICAL MECHANICS: STATICS AND THE DYNAMICS OF A PARTICLE, W. D. MacMillan. Used for over 3 decades as a self-contained and extremely comprehensive advanced undergraduate text in mathematical physics, physics, astronomy, and deeper foundations of engineering. Early sections require only a knowledge of geometry; later, a working knowledge of calculus. Hundreds of basic problems, including projectiles to the moon, escape velocity, harmonic motion, ballistics, falling bodies, transmission of power, stress and strain, elasticity, astronomical problems. 340 practice problems plus many fully worked out examples make it possible to test and extend principles developed in the text. 200 figures. xvii + 430pp. 5⅜ x 8.   S467 Paperbound $2.00

THEORETICAL MECHANICS: THE THEORY OF THE POTENTIAL, W. D. MacMillan. A comprehensive, well balanced presentation of potential theory, serving both as an introduction and a reference work with regard to specific problems, for physicists and mathematicians. No prior knowledge of integral relations is assumed, and all mathematical material is developed as it becomes necessary. Includes: Attraction of Finite Bodies; Newtonian Potential Function; Vector Fields, Green and Gauss Theorems; Attractions of Surfaces and Lines; Surface Distribution of Matter; Two-Layer Surfaces; Spherical Harmonics; Ellipsoidal Harmonics; etc. "The great number of particular cases . . . should make the book valuable to geophysicists and others actively engaged in practical applications of the potential theory," Review of Scientific Instruments. Index. Bibliography. xiii + 469pp. 5⅜ x 8.   S486 Paperbound $2.25

THEORETICAL MECHANICS: DYNAMICS OF RIGID BODIES, W. D. MacMillan. Theory of dynamics of a rigid body is developed, using both the geometrical and analytical methods of instruction. Begins with exposition of algebra of vectors, it goes through momentum principles, motion in space, use of differential equations and infinite series to solve more sophisticated dynamics problems. Partial contents: moments of inertia, systems of free particles, motion parallel to a fixed plane, rolling motion, method of periodic solutions, much more. 82 figs. 199 problems. Bibliography. Indexes. xii + 476pp. 5⅜ x 8.   S641 Paperbound $2.00

MATHEMATICAL FOUNDATIONS OF STATISTICAL MECHANICS, A. I. Khinchin. Offering a precise and rigorous formulation of problems, this book supplies a thorough and up-to-date exposition. It provides analytical tools needed to replace cumbersome concepts, and furnishes for the first time a logical step-by-step introduction to the subject. Partial contents: geometry & kinematics of the phase space, ergodic problem, reduction to theory of probability, application of central limit problem, ideal monatomic gas, foundation of thermo-dynamics, dispersion and distribution of sum functions. Key to notations. Index. viii + 179pp. 5⅜ x 8.
S147 Paperbound $1.50

ELEMENTARY PRINCIPLES IN STATISTICAL MECHANICS, J. W. Gibbs. Last work of the great Yale mathematical physicist, still one of the most fundamental treatments available for advanced students and workers in the field. Covers the basic principle of conservation of probability of phase, theory of errors in the calculated phases of a system, the contributions of Clausius, Maxwell, Boltzmann, and Gibbs himself, and much more. Includes valuable comparison of statistical mechanics with thermodynamics: Carnot's cycle, mechanical definitions of entropy, etc. xvi + 208pp. 5⅜ x 8.   S707 Paperbound $1.45

PRINCIPLES OF MECHANICS AND DYNAMICS, Sir William Thomson (Lord Kelvin) and Peter Guthrie Tait. The principles and theories of fundamental branches of classical physics explained by two of the greatest physicists of all time. A broad survey of mechanics, with material on hydrodynamics, elasticity, potential theory, and what is now standard mechanics. Thorough and detailed coverage, with many examples, derivations, and topics not included in more recent studies. Only a knowledge of calculus is needed to work through this book. Vol. I (Preliminary): Kinematics; Dynamical Laws and Principles; Experience (observation, experimentation, formation of hypotheses, scientific method); Measures and Instruments; Continuous Calculating Machines. Vol. II (Abstract Dynamics): Statics of a Particle—Attraction; Statics of Solids and Fluids. Formerly Titled "Treatise on Natural Philosophy." Unabridged reprint of revised edition. Index. 168 diagrams. Total of xlii + 1035pp. 5⅜ x 8½.
Vol. I: S966 Paperbound **$2.35**
Vol. II: S967 Paperbound **$2.35**
Two volume Set Paperbound **$4.70**

INVESTIGATIONS ON THE THEORY OF THE BROWNIAN MOVEMENT, Albert Einstein. Reprints from rare European journals. 5 basic papers, including the Elementary Theory of the Brownian Movement, written at the request of Lorentz to provide a simple explanation. Translated by A. D. Cowper. Annotated, edited by R. Fürth. 33pp. of notes elucidate, give history of previous investigations. Author, subject indexes. 62 footnotes. 124pp. 5⅜ x 8.
S304 Paperbound **$1.25**

MECHANICS VIA THE CALCULUS, P. W. Norris, W. S. Legge. Covers almost everything, from linear motion to vector analysis: equations determining motion, linear methods, compounding of simple harmonic motions, Newton's laws of motion, Hooke's law, the simple pendulum, motion of a particle in 1 plane, centers of gravity, virtual work, friction, kinetic energy of rotating bodies, equilibrium of strings, hydrostatics, sheering stresses, elasticity, etc. 550 problems. 3rd revised edition. xii + 367pp. 6 x 9.
S207 Clothbound **$4.95**

THE DYNAMICS OF PARTICLES AND OF RIGID, ELASTIC, AND FLUID BODIES; BEING LECTURES ON MATHEMATICAL PHYSICS, A. G. Webster. The reissuing of this classic fills the need for a comprehensive work on dynamics. A wide range of topics is covered in unusually great depth, applying ordinary and partial differential equations. Part I considers laws of motion and methods applicable to systems of all sorts; oscillation, resonance, cyclic systems, etc. Part 2 is a detailed study of the dynamics of rigid bodies. Part 3 introduces the theory of potential; stress and strain, Newtonian potential functions, gyrostatics, wave and vortex motion, etc. Further contents: Kinematics of a point; Lagrange's equations; Hamilton's principle; Systems of vectors; Statics and dynamics of deformable bodies; much more, not easily found together in one volume. Unabridged reprinting of 2nd edition. 20 pages of notes on differential equations and the higher analysis. 203 illustrations. Selected bibliography. Index. xi + 588pp. 5⅜ x 8.
S522 Paperbound **$2.45**

A TREATISE ON DYNAMICS OF A PARTICLE, E. J. Routh. Elementary text on dynamics for beginning mathematics or physics student. Unusually detailed treatment from elementary definitions to motion in 3 dimensions, emphasizing concrete aspects. Much unique material important in recent applications. Covers impulsive forces, rectilinear and constrained motion in 2 dimensions, harmonic and parabolic motion, degrees of freedom, closed orbits, the conical pendulum, the principle of least action, Jacobi's method, and much more. Index. 559 problems, many fully worked out, incorporated into text. xiii + 418pp. 5⅜ x 8.
S696 Paperbound **$2.25**

DYNAMICS OF A SYSTEM OF RIGID BODIES (Elementary Section), E. J. Routh. Revised 7th edition of this standard reference. This volume covers the dynamical principles of the subject, and its more elementary applications: finding moments of inertia by integration, foci of inertia, d'Alembert's principle, impulsive forces, motion in 2 and 3 dimensions, Lagrange's equations, relative indicatrix, Euler's theorem, large tautochronous motions, etc. Index. 55 figures. Scores of problems. xv + 443pp. 5⅜ x 8.
S664 Paperbound **$2.50**

DYNAMICS OF A SYSTEM OF RIGID BODIES (Advanced Section), E. J. Routh. Revised 6th edition of a classic reference aid. Much of its material remains unique. Partial contents: moving axes, relative motion, oscillations about equilibrium, motion. Motion of a body under no forces, any forces. Nature of motion given by linear equations and conditions of stability. Free, forced vibrations, constants of integration, calculus of finite differences, variations, precession and nutation, motion of the moon, motion of string, chain, membranes. 64 figures. 498pp. 5⅜ x 8.
S229 Paperbound **$2.45**

DYNAMICAL THEORY OF GASES, James Jeans. Divided into mathematical and physical chapters for the convenience of those not expert in mathematics, this volume discusses the mathematical theory of gas in a steady state, thermodynamics, Boltzmann and Maxwell, kinetic theory, quantum theory, exponentials, etc. 4th enlarged edition, with new material on quantum theory, quantum dynamics, etc. Indexes. 28 figures. 444pp. 6⅛ x 9¼.
S136 Paperbound **$2.65**

THE THEORY OF HEAT RADIATION, Max Planck. A pioneering work in thermodynamics, providing basis for most later work, Nobel laureate Planck writes on Deductions from Electrodynamics and Thermodynamics, Entropy and Probability, Irreversible Radiation Processes, etc. Starts with simple experimental laws of optics, advances to problems of spectral distribution of energy and irreversibility. Bibliography. 7 illustrations. xiv + 224pp. 5⅜ x 8.
S546 Paperbound **$1.75**

Catalogue of Dover Books

THE PRINCIPLE OF RELATIVITY, A. Einstein, H. Lorentz, H. Minkowski, H. Weyl. These are the 11 basic papers that founded the general and special theories of relativity, all translated into English. Two papers by Lorentz on the Michelson experiment, electromagnetic phenomena. Minkowski's SPACE & TIME, and Weyl's GRAVITATION & ELECTRICITY. 7 epoch-making papers by Einstein: ELECTROMAGNETICS OF MOVING BODIES, INFLUENCE OF GRAVITATION IN PROPAGATION OF LIGHT, COSMOLOGICAL CONSIDERATIONS, GENERAL THEORY, and 3 others. 7 diagrams. Special notes by A. Sommerfeld. 224pp. 5⅜ x 8.
S81 Paperbound **$1.75**

EINSTEIN'S THEORY OF RELATIVITY, Max Born. Revised edition prepared with the collaboration of Gunther Leibfried and Walter Biem. Steering a middle course between superficial popularizations and complex analyses, a Nobel laureate explains Einstein's theories clearly and with special insight. Easily followed by the layman with a knowledge of high school mathematics, the book has been thoroughly revised and extended to modernize those sections of the well-known original edition which are now out of date. After a comprehensive review of classical physics, Born's discussion of special and general theories of relativity covers such topics as simultaneity, kinematics, Einstein's mechanics and dynamics, relativity of arbitrary motions, the geometry of curved surfaces, the space-time continuum, and many others. Index. Illustrations, vii + 376pp. 5⅜ x 8.
S769 Paperbound **$2.00**

ATOMS, MOLECULES AND QUANTA, Arthur E. Ruark and Harold C. Urey. Revised (1963) and corrected edition of a work that has been a favorite with physics students and teachers for more than 30 years. No other work offers the same combination of atomic structure and molecular physics and of experiment and theory. The first 14 chapters deal with the origins and major experimental data of quantum theory and with the development of conceptions of atomic and molecular structure prior to the new mechanics. These sections provide a thorough introduction to atomic and molecular theory, and are presented lucidly and as simply as possible. The six subsequent chapters are devoted to the laws and basic ideas of quantum mechanics: Wave Mechanics, Hydrogenic Atoms in Wave Mechanics, Matrix Mechanics, General Theory of Quantum Dynamics, etc. For advanced college and graduate students in physics. Revised, corrected republication of original edition, with supplementary notes by the authors. New preface by the authors. 9 appendices. General reference list. Indices. 228 figures. 71 tables. Bibliographical material in notes, etc. Total of xxiii + 810pp. 5⅜ x 8⅜.
S1106 Vol. I Paperbound **$2.50**
S1107 Vol. II Paperbound **$2.50**
Two volume set Paperbound **$5.00**

WAVE MECHANICS AND ITS APPLICATIONS, N. F. Mott and I. N. Sneddon. A comprehensive introduction to the theory of quantum mechanics; not a rigorous mathematical exposition it progresses, instead, in accordance with the physical problems considered. Many topics difficult to find at the elementary level are discussed in this book. Includes such matters as: the wave nature of matter, the wave equation of Schrödinger, the concept of stationary states, properties of the wave functions, effect of a magnetic field on the energy levels of atoms, electronic spin, two-body problem, theory of solids, cohesive forces in ionic crystals, collision problems, interaction of radiation with matter, relativistic quantum mechanics, etc. All are treated both physically and mathematically. 68 illustrations. 11 tables. Indexes. xii + 393pp. 5⅜ x 8½.
S1070 Paperbound **$2.25**

BASIC METHODS IN TRANSFER PROBLEMS, V. Kourganoff, Professor of Astrophysics, U. of Paris. A coherent digest of all the known methods which can be used for approximate or exact solutions of transfer problems. All methods demonstrated on one particular problem —Milne's problem for a plane parallel medium. Three main sections: fundamental concepts (the radiation field and its interaction with matter, the absorption and emission coefficients, etc.); different methods by which transfer problems can be attacked; and a more general problem—the non-grey case of Milne's problem. Much new material, drawing upon declassified atomic energy reports and data from the USSR. Entirely understandable to the student with a reasonable knowledge of analysis. Unabridged, revised reprinting. New preface by the author. Index. Bibliography. 2 appendices. xv + 281pp. 5⅜ x 8½.
S1074 Paperbound **$2.00**

PRINCIPLES OF QUANTUM MECHANICS, W. V. Houston. Enables student with working knowledge of elementary mathematical physics to develop facility in use of quantum mechanics, understand published work in field. Formulates quantum mechanics in terms of Schroedinger's wave mechanics. Studies evidence for quantum theory, for inadequacy of classical mechanics, 2 postulates of quantum mechanics; numerous important, fruitful applications of quantum mechanics in spectroscopy, collision problems, electrons in solids; other topics. "One of the most rewarding features . . . is the interlacing of problems with text," Amer. J. of Physics. Corrected edition. 21 illus. Index. 296pp. 5⅜ x 8. S524 Paperbound **$2.00**

PHYSICAL PRINCIPLES OF THE QUANTUM THEORY, Werner Heisenberg. A Nobel laureate discusses quantum theory; Heisenberg's own work, Compton, Schroedinger, Wilson, Einstein, many others. Written for physicists, chemists who are not specialists in quantum theory, only elementary formulae are considered in the text; there is a mathematical appendix for specialists. Profound without sacrifice of clarity. Translated by C. Eckart, F. Hoyt. 18 figures. 192pp. 5⅜ x 8.
S113 Paperbound **$1.25**

SELECTED PAPERS ON QUANTUM ELECTRODYNAMICS, edited by **J. Schwinger.** Facsimiles of papers which established quantum electrodynamics, from initial successes through today's position as part of the larger theory of elementary particles. First book publication in any language of these collected papers of Bethe, Bloch, Dirac, Dyson, Fermi, Feynman, Heisenberg, Kusch, Lamb, Oppenheimer, Pauli, Schwinger, Tomonoga, Weisskopf, Wigner, etc. 34 papers in all, 29 in English, 1 in French, 3 in German, 1 in Italian. Preface and historical commentary by the editor, xvii + 423pp. 6⅛ x 9¼. S444 Paperbound **$2.75**

THE FUNDAMENTAL PRINCIPLES OF QUANTUM MECHANICS, WITH ELEMENTARY APPLICATIONS, E. C. Kemble. An inductive presentation, for the graduate student or specialist in some other branch of physics. Assumes some acquaintance with advanced math; apparatus necessary beyond differential equations and advanced calculus is developed as needed. Although a general exposition of principles, hundreds of individual problems are fully treated, with applications of theory being interwoven with development of the mathematical structure. The author is the Professor of Physics at Harvard Univ. "This excellent book would be of great value to every student . . . a rigorous and detailed mathematical discussion of all of the principal quantum-mechanical methods . . . has succeeded in keeping his presentations clear and understandable," Dr. Linus Pauling, J. of the American Chemical Society. Appendices: calculus of variations, math. notes, etc. Indexes. 611pp. 5⅜ x 8. S472 Paperbound **$3.00**

QUANTUM MECHANICS, H. A. Kramers. A superb, up-to-date exposition, covering the most important concepts of quantum theory in exceptionally lucid fashion. 1st half of book shows how the classical mechanics of point particles can be generalized into a consistent quantum mechanics. These 5 chapters constitute a thorough introduction to the foundations of quantum theory. Part II deals with those extensions needed for the application of the theory to problems of atomic and molecular structure. Covers electron spin, the Exclusion Principle, electromagnetic radiation, etc. "This is a book that all who study quantum theory will want to read," J. Polkinghorne, PHYSICS TODAY. Translated by D. ter Haar. Prefaces, introduction. Glossary of symbols. 14 figures. Index. xvi + 496pp. 5⅜ x 8⅜. S1150 Paperbound **$2.75**

THE THEORY AND THE PROPERTIES OF METALS AND ALLOYS, N. F. Mott, H. Jones. Quantum methods used to develop mathematical models which show interrelationship of basic chemical phenomena with crystal structure, magnetic susceptibility, electrical, optical properties. Examines thermal properties of crystal lattice, electron motion in applied field, cohesion, electrical resistance, noble metals, para-, dia-, and ferromagnetism, etc. "Exposition . . . clear . . . mathematical treatment . . . simple," Nature. 138 figures. Bibliography. Index. xiii + 320pp. 5⅜ x 8. S456 Paperbound **$2.00**

FOUNDATIONS OF NUCLEAR PHYSICS, edited by **R. T. Beyer.** 13 of the most important papers on nuclear physics reproduced in facsimile in the original languages of their authors: the papers most often cited in footnotes, bibliographies. Anderson, Curie, Joliot, Chadwick, Fermi, Lawrence, Cockcroft, Hahn, Yukawa. UNPARALLELED BIBLIOGRAPHY. 122 double-columned pages, over 4,000 articles, books, classified. 57 figures. 288pp. 6⅛ x 9¼. S19 Paperbound **$2.00**

MESON PHYSICS, R. E. Marshak. Traces the basic theory, and explicitly presents results of experiments with particular emphasis on theoretical significance. Phenomena involving mesons as virtual transitions are avoided, eliminating some of the least satisfactory predictions of meson theory. Includes production and study of π mesons at nonrelativistic nucleon energies, contrasts between π and μ mesons, phenomena associated with nuclear interaction of π mesons, etc. Presents early evidence for new classes of particles and indicates theoretical difficulties created by discovery of heavy mesons and hyperons. Name and subject indices. Unabridged reprint. viii + 378pp. 5⅜ x 8. S500 Paperbound **$1.95**

Prices subject to change without notice.

Dover publishes books on art, music, philosophy, literature, languages, history, social sciences, psychology, handcrafts, orientalia, puzzles and entertainments, chess, pets and gardens, books explaining science, intermediate and higher mathematics, mathematical physics, engineering, biological sciences, earth sciences, classics of science, etc. Write to:

Dept. catrr.
Dover Publications, Inc.
180 Varick Street, N.Y. 14, N.Y.